Physically Unclonable Functions

Roel Maes

Physically Unclonable Functions

Constructions, Properties and Applications

 Springer

Roel Maes
ESAT-COSIC
Dept. of Electrical Engineering
KU Leuven
Heverlee, Belgium

ISBN 978-3-662-51452-8   ISBN 978-3-642-41395-7 (eBook)
DOI 10.1007/978-3-642-41395-7
Springer Heidelberg New York Dordrecht London

Printed on acid-free paper

Springer is part of Springer Science+Business Media (www.springer.com)

To my wife, Sofie, without whom I would not be half the man I am today, and to my daughter, Mira, the greatest joy in my life.

Preface

Information security techniques are indispensable in a world that relies, in a continuously increasing extent, on digital information processing and communication systems. The use of cryptographic primitives enables us to reduce information security goals to physical security requirements, such as secure algorithm execution, and the secure generation and storage of secrets. *Physically unclonable functions*, or *PUFs*, are innovative physical security primitives which produce unclonable and inherent instance-specific measurements of physical objects; PUFs are in many ways the inanimate equivalent of biometrics for human beings. Since they are able to securely generate and store secrets, PUFs allow us to bootstrap the physical implementation of an information security system. In this book, we discuss PUFs in all their facets: the multitude of their physical *constructions*, the algorithmic and physical *properties* which describe them, and the techniques required to deploy them in security *applications*.

We first give an extensive overview and classification of PUF constructions, with a focus on *intrinsic PUFs*. We identify significant subclasses, implementation properties and general design techniques used to amplify sub-microscopic physical distinctions into observable digital response vectors. We list the useful properties attributed to PUFs and capture them in descriptive yet clear definitions. Through an elaborate comparative analysis, PUF-defining properties are distinguished from nice-to-have but not strictly required qualities. Additionally, a formal framework for deploying PUFs and similar physical primitives in cryptographic reductions is described.

In order to objectively compare the quality of different PUF constructions, we contributed to the development of a silicon test platform carrying six different intrinsic PUF structures. Based on experimental data from 192 distinct test devices, including measurements at temperature and supply voltage corner cases, the reliability, the uniqueness and the unpredictability of each of these constructions is assessed, and summarized in concise yet meaningful statistics.

Their instance-specific and unclonable nature enables us to use PUFs as entity identifiers. In combination with appropriate processing algorithms, they can even authenticate entities and securely generate and store secrets. We present techniques to achieve PUF-based entity identification, entity authentication, and secure key generation. Additionally, we propose practical designs that implement these techniques, and derive and calculate meaningful measures for assessing the performance of different PUF constructions in these applications, based on the quality of their response statistics. Finally, as a proof of concept, we present a fully functional prototype implementation of a PUF-based cryptographic key generator, demonstrating the full benefit of using PUFs and the efficiency of the introduced processing techniques.

Acknowledgements

The origin of this book is the thesis I wrote as the conclusion of my Ph.D. degree, discussing the main research topics I worked on, and contributed to, between 2007 and 2012. It is evident that a Ph.D. can only be successfully completed with the help of many people and instances.

Before all others, I am grateful to my promoter Prof. Ingrid Verbauwhede, for the opportunities she created for me, for her advice on matters great and small, and for trusting me to find my own way. I am also much indebted to Dr. Pim Tuyls for introducing me to the exciting topic of this book, for guiding me through my first couple of years as a young researcher, and in general for taking the idea of PUFs to a whole new level. In addition, I want to thank the other members of my Ph.D. jury for their combined effort in reviewing the text of my thesis, the KU Leuven for offering such an inspiring academic environment, and the Agency for Innovation by Science and Technology (IWT) for funding the major part of my Ph.D. research.

Being an academic researcher is far from a solitary occupation, and I have had the pleasure and privilege of meeting and collaborating with many of my peers worldwide. I must acknowledge all of my appreciated coauthors over the years, for their guidance and contribution. A special thanks goes out to the partners of the European UNIQUE project, and the people behind them, with whom it was always a pleasure to meet and discuss things, and from whom I have learned a lot. I also greatly enjoyed the opportunities to get a taste of life as a researcher in industry, through internships at Philips and Intel; both were extremely challenging experiences which have had a significant positive impact on me.

It is hard to overestimate my gratitude for having been able to work for five years in an atmosphere as vibrant, yet warm and friendly as the COSIC research group. Over the years I have seen people come and go, but the helpfulness, sociability and plain fun were invariably present. One of the many special people responsible for this is Péla Noé, COSIC's secretary and so much more, but I am grateful to all my colleagues who made COSIC such an enjoyable place to work.

Geel, Belgium   Roel Maes
September 2013

Contents

Abbreviations

AES	Advanced Encryption Standard
ALILE	Aluminum-Induced Layer Exchange
ANN	Artificial Neural Network
ASIC	Application-Specific Integrated Circuit
BCH	Bose, Chaudhuri and Hocquenghem (code)
CD	Compact Disc
CETS	Commission on Engineering and Technical Systems
CMOS	Complementary Metal-Oxide-Semiconductor
COTS	Commercial off-the Shelf
CPUF	Controlled PUF
DNA	Deoxyribonucleic Acid
FF	Feed-Forward
FIB	Focused Ion Beam
FPGA	Field-Programmable Gate Array
GMC	Generalized Multiple Concatenated (decoding algorithm)
HH	High Temperature/High Supply Voltage (condition)
HL	High Temperature/Low Supply Voltage (condition)
IC	Integrated Circuit
ICID	IC Identification
IP	Intellectual Property
L.G.	Lehmer-Gray
LC	Inductor(L)-Capacitor(C)
LDPC	Low-Density Parity-Check
LFSR	Linear Feedback Shift Register
LH	Low Temperature/High Supply Voltage (condition)
LL	Low Temperature/Low Supply Voltage (condition)
LQFP	Low-Profile Quad Flat Package
LRPUF	Logically Reconfigurable PUF
MAC	Message Authentication Code
μC	Microcontroller
MOSFET	Metal-Oxide-Semiconductor Field-Effect Transistor

MS	Mixed-Signal
mux	Multiplexer
NAND	Not AND (logical)
n-MOS	n-Channel MOSFET
NOR	Not OR (logical)
P.C.	Pairwise Comparison
PCA	Principal Component Analysis
p-MOS	p-Channel MOSFET
POK	Physically Obfuscated Key
POWF	Physical One-Way Function
PPUF	Public PUF
PRF	Pseudo-random Function
PRNG	Pseudo-random Number Generator
PUF	Physically Unclonable Function
R/W	Read/Write (interface)
RAM	Random-Access Memory
REF	Reference (condition)
REP	Repetition (code)
RF	Radio Frequency
RFID	Radio Frequency Identification
ROC	Receiver-Operating Characteristic
ROM	Read-Only Memory
ROPUF	Ring Oscillator PUF
RPUF	Reconfigurable PUF
RSA	Rivest, Shamir and Adleman (algorithm)
RTL	Register Transfer Level
SDML	Soft-Decision Maximum-Likelihood (decoding algorithm)
SHA	Secure Hash Algorithm
SHIC	Super-High Information Content
SIMPL	Simulation Possible but Laborious
SNM	Static Noise Margin
SPI	Serial Peripheral Interface
SR	Set/Reset
SRAM	Static Random-Access Memory
SVM	Support Vector Machine
TRNG	True Random Number Generator
TSMC	Taiwan Semiconductor Manufacturing Company Limited
VHDL	VLSI Hardware Description Language
VLSI	Very-Large-Scale Integration
WC	Worst Case (condition)
XOR	Exclusive OR (logical)

List of Figures

List of Tables

Chapter 1
Introduction and Preview

1.1 Introduction

1.1.1 Trust and Security in a Modern World

Trust is a sociological concept expressing the positive belief that a person or a system we interact with will behave as expected. In our day-to-day life, we constantly and often implicitly put our trust in other parties, e.g.:

- When we drive a car, we trust that the car will function as expected, that the brakes will work and that the car goes right when we turn the steering wheel right. We also trust that the other people driving cars around us are qualified to drive a car and are paying attention to traffic.
- When we deposit our money in a bank account, we trust that the bank will keep the money safe.
- When we send someone a letter, we trust the postal services to deliver the letter in a timely manner to the right person, and to keep the letter closed such that no one else can read its content.
- When we buy something in a shop, we trust the shop owner to deliver the product, e.g. when we pay in advance, and that we receive the genuine product we paid for. On the other hand, the shop owner trusts that we will pay for all products we carry out.

In the majority of situations, such trust-based interactions work out in the right way, because the parties we interact with are *trustworthy*. In fact, our entire complex society is based on such trust relations between people and systems, and it would not last very long if no one or no thing could be trusted.

However, we don't live in an ideal world, and it would be very naive to think that everyone is intrinsically trustworthy. Many parties have external motives to behave in a trustworthy manner, e.g. the shop and the bank won't get many customers when they cannot be trusted, and the other car owners will primarily drive carefully for their own safety. Some parties cannot be trusted at all; we immediately think of

R. Maes, *Physically Unclonable Functions*, DOI 10.1007/978-3-642-41395-7_1,
© Springer-Verlag Berlin Heidelberg 2013

criminals and terrorists, but this can also include e.g., disgruntled employees, envious colleagues or nosy neighbors, or even normally honest people who are tempted to abuse a situation when it presents itself. We need systems that induce, guarantee or even enforce trustworthiness of parties in our non-ideal world. This is what we call *security*, i.e. security is a means to enable trust.

In the past, and to a large extent still today, security is either based on physical protection and prevention measures, on observation and detection of untrusted elements, or on legal and other reprimands of trust violations, and often on a combination of these techniques. For example, in order to keep its (your) money secure, a bank will store it in a vault (physical protection). The access to this vault is moreover strictly limited to the bank's employees and protocols are in place to keep other people away (detection). Finally, by law, trying to rob a bank is also a criminal act for which one will be prosecuted if caught (legal reprimands). In our rapidly digitalizing modern world, these security techniques are by themselves often no longer sufficient to adequately enable trusted interactions, both due to (i) the nature of these interactions, and (ii) the scale of the possible threats.

(i) The remote and generic nature of many digital interactions lacks physical protection and assurance measures, many of which are even implicitly present in non-digital communications. For example, in the past, most interactions with your bank would take place inside the bank's building, face-to-face with one of the bank's employees. You (implicitly) trusted the authenticity of this interaction, e.g. because the building was always in the same place, and perhaps because you physically recognized the clerk from previous transactions, and vice versa. However, in the last couple of years, interactions with your bank have shifted largely to online banking systems. In such an online system, e.g. a website, this implied notion of authenticity no longer exists, since anyone could set up a website resembling that of your bank, and even fake its web address. The same holds from the bank's perspective: anyone could log in to the website and claim to be you. Other security measures are needed to guarantee the authenticity of this interaction.

(ii) The main success of digitalization is that it enables automation of information processes to very large scales and speeds. However, this is also one of the main risk factors when it comes to digital crime. For example, in the real (non-digital) world, there is a risk of having your wallet stolen on the street. However, a thief will have to focus on one victim at a time, and for each attempt there exists a significant risk of failure which often ends in getting caught. In a vastly interconnected computer network like the Internet, with hundreds of millions of simultaneously active users, a *digital* thief can deploy a computer program which targets thousands or millions at a time at an incredibly fast pace. Moreover, failed attacks typically go by unnoticed or are hard to trace back, and even with a very small success rate the thief will get a significant return due to the vast number of targeted victims. Like the threat, the security measures will also need to be digitized and automated in order to offer adequate protection.

1.1.2 Information Security and Cryptology

Information Security

Information security deals with securing interactions involving the communication of information. The need for information security has existed for a long time, historically in particular for matters of love and hate, i.e. secret love letters and sensitive warfare communication. However, in the last couple of decades this need has risen exponentially due to our vast and ever increasing reliance on digital information processing and communication systems. Unimaginable quantities of private and often sensitive information are stored and communicated over the Internet and other digital networks every second. Through the pervasiveness of smart mobile personal devices, digital technology impacts our daily lives in ways we could not have foreseen, and with the introduction and tremendous success of social networks, it has even become an integral part of our lives. In many ways our society has become a flow of digital information, and reliable information security techniques are indispensable to enable trust in this digital world.

Information security techniques are most comprehensibly classified by means of the goals they aim to achieve. The most important goals are:

- *Data confidentiality* relates to keeping information secret from unauthorized parties, e.g. when accessing your bank account statements online, you don't want anyone else to see this information.
- *Entity authentication* deals with obtaining proof of the identity and the presence of the entity one is interacting with, e.g. in an online banking system, you need proof that you're dealing with the real website of your bank, and your bank needs proof that you are who you claim to be before granting access to your account.
- *Data integrity and authentication* is aimed at preventing and detecting unauthorized alteration of data (integrity) and ensuring the origin of the data (authentication), e.g. when you issue an online bank transfer, your bank needs to be sure that it was you who issued the transfer, and that the data of the transfer (amount, account number, . . .) has not been changed by someone who could have intercepted the transfer message before it reached the bank.

Cryptology

Cryptography, a subfield of cryptology, deals with the construction of protocols and algorithms to achieve information security goals, typically on a mathematical basis. The other subfield of cryptology is cryptanalysis, which analyzes the security of cryptographic constructions by attempting to *break* their anticipated security. Both subfields are intimately linked, often exercised by the same persons, and a close interplay between both is invaluable. One of the, if not *the* basic principle of modern cryptology is the understanding that a cryptographic construction can only be considered secure if its internal workings are general knowledge and have successfully withstood elaborate cryptanalysis attempts from independent parties. This is also called *Kerckhoffs' principle* after Auguste Kerckhoffs who first stated it [66], and stands in contrast to so-called *security-through-obscurity* which attempts to reach security goals with undisclosed and hence unanalyzed constructions.

A basic design principle for many cryptographic constructions is to reduce the security goal they attempt to achieve to the secrecy of a single parameter in the construction, called the *key*. The obtained level of security is typically expressed by the required effort to break it without knowing the key, which should be an exponential function of the key's length in bits. An important aspect in these security reductions is the assumptions one makes about the power of the adversary, e.g. whether he can observe a number of inputs and/or outputs of a primitive and whether he is only a passive observer or if he can actively or even adaptively interfere with the execution of the primitive. Based on the nature of a reduction, different security notions can be distinguished:

- *Heuristic security* means that even after elaborate cryptanalysis of a construction, no attacks can be found which break its security with a computational effort less than expressed by the key length.
- *Provable security* means that, through logical reasoning, the construction's security can be shown to be equivalent to a mathematical problem which is perceived to be *hard*, with the problem's hardness expressed by the key length. Examples of such hard mathematical problems for which no efficient algorithms are known, and which are actually used in cryptographic constructions, are factorization of large integers (e.g. as used in the RSA algorithm [111]) and computation of discrete logarithms (e.g. as used in the Diffie-Hellman key exchange protocol [31]).
- *Information-theoretical security* means that it can be shown through information-theoretical reasoning that an adversary does not have sufficient information to break the construction's security. This basically means the construction is unbreakable, even to an adversary with unlimited computational capabilities.

For an extensive overview of the construction and properties of cryptographic primitives, we refer to [96]. Cryptographic primitives can be classified based on the nature of their key. We respectively distinguish (i) unkeyed primitives, (ii) symmetric-key primitives, and (iii) public-key primitives and list some of their most important instantiations and achieved security goals.

(i) Unkeyed primitives are constructions which do not require a key. Following Kerckhoffs' principle, their operation is hence completely public and can be executed by everyone. Their security is basically grounded in the difficulty of finding an input which matches a given output. The most used unkeyed primitives are cryptographic hash functions, which provide data integrity and also often serve as a building block in larger cryptographic constructions.

(ii) Symmetric-key primitives are based on a single key which is only known to authorized parties and secret to anyone else. Symmetric-key encryption algorithms, such as block ciphers and stream ciphers, provide data confidentiality between parties knowing the secret key. Symmetric-key message authentication codes provide data integrity and authentication and entity authentication between parties knowing the key.

(iii) Public-key primitives are based on a key pair, one of which is public and the other is kept private. In a public-key encryption scheme, everyone can encrypt a message with the public key, but only the party which knows the private key can decrypt it. In a public-key signature scheme, only the party knowing the private key can generate a signature on a message, and everyone can use the public key to verify that party's signature. Signature schemes provide entity authentication, among other goals.

1.1.3 Physical Security and Roots of Trust

Physical Security

To use a cryptographic primitive in practice, it needs to be implemented on a digital platform in an efficient manner. Unlike Kerckhoffs' principle for the general construction, for the implementation it is typically assumed that the primitive behaves like a *black box*, i.e. one is only able to observe the input-output behavior of the implementation, not its internal operations. In particular, for nearly all (keyed) cryptographic primitives, it is assumed that:

- A secure (random, unique, unpredictable, . . .) key can be generated for every instantiation of the primitive. This is called *secure key generation*.
- The key can be assigned to, stored and retrieved by the instantiation without being revealed. This is called *secure key storage*.
- The instantiation can execute the cryptographic algorithm without revealing any (partial) information about the key or about internal results, and without an outsider being able to influence the internal execution in any possible way. This is called *secure execution*.

While these are convenient assumptions for mathematical security reductions, from a practical perspective they are very hard to attain. Moreover, it is clear that none of these three black-box assumptions can be achieved through information security techniques, but require physical security measures. In a way, one could say that cryptographic primitives reduce information security objectives into physical security requirements.

The fact that none of the three identified physical security objectives are trivial is made clear by the numerous cases where information security systems are attacked by breaking the security at a physical level.

- The fact that secure key generation is difficult was just recently made clear again by Lenstra et al. [77], who show that there is a significant shortage in randomness in a large collected set of actually used public keys from a public key signature scheme, likely caused by badly implemented key generators. For some of the keys in the analyzed collection, this leads to an immediate loss of security.

- Storing secret keys in a highly secure manner partially contradicts the fact that they still need to be in some (permanent) digital format to be usable in an algorithm. For typical digital implementations, this means that the key bits reside somewhere in a non-volatile digital memory on a silicon chip. Even with extensive countermeasures in place, it is very difficult to stop a well-equipped and/or determined adversary from gaining physical access to key memories, e.g. as demonstrated by Torrance and James [143] and Tarnovsky [138].
- There are many ways an adversary can break the secure execution assumption, both on the software and on the hardware level. Modern cryptographic implementations can no longer ignore side-channel attacks, which abuse the fact that all actions on a digital platform leak information about their execution through so-called side channels, e.g. through their execution time [68], their power consumption [69], their electro-magnetic radiation [107], etc. Fault attacks [10] on the other hand seek to disrupt the expected execution of a cryptographic algorithm through physical means, and learn sensitive information from the faulty results.

Physical Roots of Trust

In order to provide these physical security objectives, we cannot rely on mathematical reductions anymore. Instead, we need to develop physical techniques and primitives which, based on physical reasoning, can be *trusted* to withstand certain physical attacks and can hence provide certain physical security objectives. We call such primitives *physical roots of trust*. Figure 1.1 shows how information security objectives can be achieved from physical security and eventually from physical roots of trust, i.e. trusted primitives which are rooted in the actual physical world. Possible candidates of physical roots of trust are:

- True random number generators or TRNGs [37, 122] harvest random numbers from truly physical sources of randomness and can therefore be trusted to produce highly random keys for cryptographic purposes.
- Design styles for digital silicon circuits have been developed which minimize and ideally eliminate certain physical side channels [141].
- Physically unclonable functions or PUFs produce unpredictable and instance-specific values and can be used to provide physically secure key generation and storage. They are the main subject of this book.

1.2 Preview

1.2.1 Introducing Physically Unclonable Functions

A physically unclonable function or PUF is best described as "an expression of an inherent and unclonable instance-specific feature of a physical object", and as such

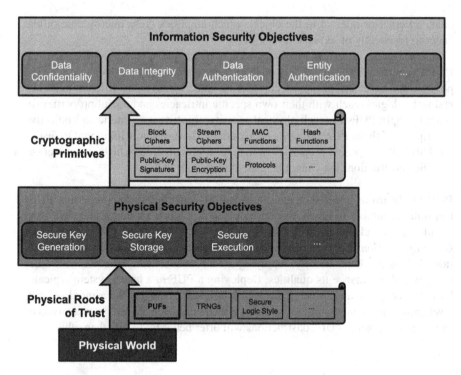

Fig. 1.1 Relations between information security, cryptography, physical security and physical roots of trust

has a strong resemblance to biometric features of human beings, like fingerprints. To be specific, PUFs show qualities which cannot be obtained from cryptographic reductions, but require a physical basis to establish them, the most noteworthy being *physical unclonability*. This means that through physical reasoning it is shown that producing a physical clone of a PUF is extremely hard or impossible.

PUF Constructions The physical motivation for claiming unclonability of an inherent instance-specific feature is found in the technical limitations of the production of physical objects. Even with extreme control over a manufacturing process, no two physically exactly identical objects can be created due to the influence of random and uncontrollable effects. Typically, these influences are very small and only take effect at (sub-)microscopic scales, but leave their random marks nonetheless. A high-precision measurement of these marks serves as an inherent and instance-specific feature. Moreover, creating a second object which produces a similar measurement is infeasible from a physical perspective, and often even technically impossible. Generating such a measurement with an accuracy high enough to distinguish these instance-specific features is the primary goal in the study of *PUF constructions*. The basic technique which is typically used is to design a construction, either external or internal to the object, which amplifies these microscopic differences to practically observable levels.

PUF Properties   A wide variety of PUF constructions based on this principle are possible and have been proposed, considering objects from many different materials and technologies, each with their own specific intricacies and useful properties. In order to apply PUFs to reach physical security objectives, a generic and objective description of these *PUF properties* is required. Moreover, it is important to distinguish truly PUF-specific properties from other useful qualities which are inherent to specific constructions but cannot be generalized to all PUFs.

PUF Applications   Based on their unclonability and other useful properties, PUFs can fulfill a number of physical security objectives when applied in the right way. Besides taking advantage of the physical security properties of PUFs, such *PUF-based applications* also need to deal with the practical limitations of the construction. This is accomplished by deploying a PUF in a scheme together with other primitives that enhance its qualities. Deploying a PUF in a larger system typically leads to trade-offs, and hence optimization problems, between the aspired security level and the implementation restrictions of the application. Based on an analysis of such a scheme, some PUF constructions will offer better trade-offs than others.

1.2.2 Book Outline

In this book, we study PUF constructions, properties as well as applications, both from a conceptual and from a very practical perspective. Figure 1.2 shows how these subjects relate to each other and are organized in this text.

In Chap. 2, we explain the details of the PUF concept and provide an extensive overview of existing PUF constructions with a focus on so-called *intrinsic* PUFs. This overview is of an unprecedented completeness and serves as a great aid in understanding the true nature of what we rather intuitively have called a PUF. Based on this overview, we also manage to identify significant subclasses, design techniques, implementation properties and even open problems related to PUF constructions.

In Chap. 3, we identify and define different meaningful properties attributed to PUFs and, based on Chap. 2, we analyze if and to what extent actual PUF constructions attain them. From this analysis, a number of these properties are found to be defining for the concept of a PUF, while others are mere convenient qualities but are in no way guaranteed for all PUFs. In order to increase their potential in theoretical constructions, a highly formal framework for using PUFs and their most important properties is also discussed.

In Chap. 4, we discuss the implementation of a significant subset of studied intrinsic PUF constructions on a realistic silicon platform (65 nm CMOS ASIC) and experimentally verify their behavior at nominal condition and at extreme temperature and voltage corners. We capture the qualities of each studied construction in

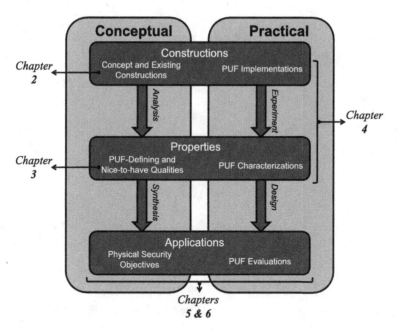

Fig. 1.2 Organization of the subjects in this book and its chapters

a small number of meaningful statistics. Additionally, we analyze the unpredictability of each PUF by introducing heuristic upper bounds on their entropy density.

In Chap. 5, we investigate how PUFs can be used to identify distinct objects, and ultimately provide entity authentication. Quality metrics for assessing identification performance are discussed and applied on the PUF constructions studied in Chap. 4, yielding a classification of their identifying capabilities. We discuss a PUF-based authentication protocol innovatively combining a PUF and other primitives. Authentication performance metrics similar to those for identification are assessed for the constructions from Chap. 4.

In Chap. 6, it is explained how PUFs can be used to obtain secure key generation and storage. Existing notions and techniques for key generation are discussed, and a practical new variant is proposed which yields a significant gain in efficiency. Based on the design constraints of a convenient construction of a practical PUF-based key generator, the PUF implementations from Chap. 4 are assessed for their key generation capacities. To conclude, we present a front-to-back PUF-based key generator design and a fully functional FPGA reference implementation thereof.

In Chap. 7, we summarize the most important aspects of this book and propose a number of interesting future research directions in this topic.

Chapter 2
Physically Unclonable Functions: Concept and Constructions

2.1 Introduction

2.1.1 The PUF Concept

Since it is the main subject of this book, it is important to clarify and define the basic concept of a physically unclonable function or PUF. However, for a variety of reasons, this task turns out to be less straightforward than expected. The collection of proposed constructions labelled 'PUF' is growing at such a pace, both in depth and breadth, that one can easily call it a *zoo* at this moment.[1] There is also an equally substantial group of constructions which could, but are for the moment not, called 'PUFs', among other reasons because they were proposed before the acronym 'PUF' had been coined, or because they were proposed by authors unfamiliar with the acronym, e.g. in fields outside hardware security engineering. All these differing constructions are moreover proposed in a greatly varying patchwork of implementation materials, technologies and platforms. Finding similar or even identifying properties which accurately capture what is understood to be a PUF is hence far from trivial. In this chapter and the next, we discuss the PUF concept in great detail in order to do exactly that: first, by extensively studying existing PUF constructions in Chap. 2, and next by describing, assessing and finally formalizing the observed properties of PUFs in Chap. 3.

On PUFs and Fingerprints

In an attempt to express the concept of a PUF in a single phrase, one of the best possible descriptions would be: "a PUF is an object's fingerprint". PUFs are similar to fingerprints for more than one reason:

---

[1] By analogy with the Complexity Zoo [139] and the SHA-3 Zoo [140].

R. Maes, *Physically Unclonable Functions*, DOI 10.1007/978-3-642-41395-7_2,
© Springer-Verlag Berlin Heidelberg 2013

- A human fingerprint is a feature which strongly expresses *individualism,* i.e. it answers the question "Which human is this?" as opposed to other features such as having two eyes and ears and walking on two legs, which express *essentialism,* i.e. answering the question "What is a human?". In a more inanimate sense, a PUF bears the same meaning for a class of objects: a PUF is an identifying feature of a specific instance of a class of objects, or for short is an *instance-specific feature.*
- As an individualising feature, a fingerprint is also *inherent,* i.e. every human being is born with fingerprints, unlike other identifying qualities like a name or a written signature, which are bestowed upon or acquired by an individual after birth. In the same way, PUFs are inherently present in an object from its creation, as a result of unique variations during its creation process.
- Finally, as an inherent individualising feature, fingerprints are also *unclonable,* i.e. the physical and biological processes which determine a human being's fingerprints are beyond any level of meaningful control which would allow us to create a second individual with the same inherent fingerprints.[2] This even holds for human beings sharing the same genetic material, like identical twins. Hence even a (hypothetical) biological clone of a person would not share that person's fingerprints. In this aspect PUFs are very similar to fingerprints. In fact, by its being the central specifier in the term 'physically unclonable functions', (physical) unclonability is one of *the* core properties of a PUF.

Following this discussion, we propose the following colloquial dictionary definition of the PUF concept: *"a PUF is an expression of an inherent and unclonable instance-specific feature of a physical object"*. By indicating that a PUF is always object-related, we explicitly distinguish PUFs from fingerprints and other biometric parameters which naturally reflect on human beings. However, as discussed above, in many aspects PUFs and biometrics are equivalent.

2.1.2 Chapter Goals

Order in the PUF Zoo

The amount of published results on physically unclonable functions has increased exponentially over the last few years, and the need for a synthesizing effort presents itself. This chapter presents an objective, large-scale and in-depth overview of the myriad of PUFs which have been proposed over the years. The main goals of this overview are:

- Present an as complete as possible reference list of PUF and PUF-like proposals to date, with a focus on so-called *intrinsic* PUFs.

---

[2]By unclonable we do not mean that it is impossible to create or obtain a facsimile of a person's fingerprints; in fact human beings create copies of their fingerprints every time they touch a smooth surface.

- Introduce and apply a classification of PUF constructions based on practical considerations.
- Present a common descriptive framework for assessing the basic functionality and quality of a PUF, and use it to produce an as fair as possible comparative quantitative analysis of different PUF implementations based on published experimental results.

Besides these clear objectives, this overview will also be of great aid in identifying interesting subclasses, properties, applications, and even open problems in the field of PUFs.

2.1.3 Chapter Overview

Before we start describing different PUF constructions and their qualities, we will in Sect. 2.2 introduce the basic nomenclature and notational conventions related to PUFs which we will use throughout this book. These conventions make it much easier to relate and compare identical concepts in often widely differing implementations and data sets, and it is therefore highly recommended you read this section first. In Sect. 2.3, we delve deeper into the semantics of the acronym 'PUF' and also discuss a number of possible classifications in the large variety of PUF constructions which have been proposed over time. The main body of this chapter is a very extensive overview of known PUF constructions,[3] detailing the implementation of each proposal and reporting experimental results, if any. In this book, we mainly focus on *intrinsic* PUF constructions which are discussed in Sect. 2.4. For more information on existing *non-intrinsic* PUF constructions we refer to Appendix B. In Sect. 2.5, a number of concepts are discussed which present extensions or modes of operation of PUFs. Finally, we conclude this chapter in Sect. 2.6.

2.2 Preliminaries

2.2.1 Conventions on Describing PUFs

Formally describing the operation of a PUF is a tedious yet important task. Due to the multitude of different PUF constructions and different ways of considering a PUF outcome, one quickly runs the risk of either introducing confusion due to an ambiguous description or losing important details due to a too sparse description. Here, we introduce the verbal and notational description of PUFs which we will use and build upon throughout the rest of this book. Since this is only intended as

---

[3] We aim at giving an exhaustive overview of PUF proposals to date, but with the plethora of new constructions appearing in recent years, it is very likely that some are missing.

a convenient but clear notation format, we refrain at this point as much as possible from making assumptions or putting restrictions on the underlying notions. Instead, we limit these conventions to the most basic concepts required to formally talk about PUFs. For completeness, the basic mathematical notations and definitions used in this section and the rest of this book are shortly introduced in Appendix A.

PUF Class

We first introduce the notion of a PUF class, denoted by $\mathcal{P}$, which is the complete description of a particular PUF construction type. A PUF class provides a creation procedure $\mathcal{P}$.Create which is invoked to create instances of $\mathcal{P}$. In general $\mathcal{P}$.Create is a probabilistic procedure, which we make explicit when necessary by providing it with a randomized input $\mathcal{P}$.Create(r^C), with $r^C \xleftarrow{\$} \{0, 1\}^*$ representing an undetermined number of fair coin tosses.

From a practical point of view, $\mathcal{P}$ represents the detailed structural design or *blueprint* of a PUF construction and $\mathcal{P}$.Create the detailed (physical) production process to build the design.

PUF Instance

A PUF instance puf is a discrete instantiation of a PUF class $\mathcal{P}$, as generated by its creation procedure $\mathcal{P}$.Create. In all respects, a PUF class can be considered to be the set of all its created instances:

$$\mathcal{P} \equiv \left\{ \mathsf{puf}_i \leftarrow \mathcal{P}.\mathsf{Create}\left(r_i^C\right) : \forall i, r_i^C \xleftarrow{\$} \{0, 1\}^* \right\}.$$

A PUF instance puf is considered as a particular crystallized *state* of the construction described by its PUF class $\mathcal{P}$. The construction of many PUF classes makes it possible to configure part of the PUF instance's state, i.e. it is not fixed by the creation procedure but can be set and easily modified by means of an external input. When required, we specify the configurable state x of a PUF instance puf as $\mathsf{puf}(x)$.

From a practical point of view, the *state* represented by a PUF instance puf is the exact (physical) structure of a produced PUF construction. The configurable part of a PUF instance's state is generally called the *challenge* which is *applied* to the PUF instance, and we will use the same terminology in this book. The set of all possible challenges x which can be applied to an instance of a PUF class $\mathcal{P}$ is denoted as $\mathcal{X}_{\mathcal{P}}$.

PUF Evaluation

Every PUF instance puf provides an evaluation procedure puf.Eval which produces a quantitative outcome representing a measurement of the PUF instance. The outcome produced by puf.Eval depends on the state represented by the PUF instance. When the PUF instance is *challengeable*, we write: $\mathsf{puf}(x).\mathsf{Eval}$. In general $\mathsf{puf}(x).\mathsf{Eval}$

is also a probabilistic procedure which we again make explicit when necessary as $\mathsf{puf}(x).\mathsf{Eval}(r^{\mathsf{E}} \xleftarrow{\$} \{0, 1\}^*)$.

From a practical point of view, a PUF instance evaluation is the outcome of a (physical) experiment, generating a particular measurement of the PUF instance's physical state at that moment. Such a measurement is generally called the *response* of the PUF instance and we will use the same terminology. The class of all possible response values which a PUF instance of a PUF class $\mathcal{P}$ can produce is denoted as $\mathcal{Y}_\mathcal{P}$.

For many PUF classes, the outcome of a PUF instance evaluation is also affected by external physical parameters, e.g. environment temperature, supply voltage level, etc. We call this the *condition* of the evaluation. When required, we denote this as $\mathsf{puf}(x).\mathsf{Eval}^\alpha$, e.g. with $\alpha = (T_{env} = 80\ °\mathrm{C})$, meaning that this evaluation took place at an environment temperature of 80 °C. When the condition specifier α is omitted, one may assume an evaluation at nominal operating conditions.

Shorthand Notation

To avoid the rather verbose use of the randomization variables r^{C} and r^{E}, we introduce a more convenient and compact notation using random variables.

Creation of a random PUF instance:

$$\mathsf{puf}_i \leftarrow \mathcal{P}.\mathsf{Create}\big(r_i^{\mathsf{C}} \xleftarrow{\$} \{0, 1\}^*\big) \text{ becomes}$$

$$\mathsf{PUF} \leftarrow \mathcal{P}.\mathsf{Create}, \text{ or the even shorter } \mathsf{PUF} \leftarrow \mathcal{P}.$$

Random evaluation of PUF instance puf_i on challenge x:

$$y_i^{(j)}(x) \leftarrow \mathsf{puf}_i(x).\mathsf{Eval}\big(r_j^{\mathsf{E}} \xleftarrow{\$} \{0, 1\}^*\big) \text{ becomes}$$

$$Y_i(x) \leftarrow \mathsf{puf}_i(x).\mathsf{Eval}, \text{ or the even shorter } Y_i(x) \leftarrow \mathsf{puf}_i(x).$$

Random evaluation of a random PUF instance on challenge x:

$$Y(x) \leftarrow \mathsf{PUF}(x).\mathsf{Eval}, \text{ or the shorter } Y(x) \leftarrow \mathsf{PUF}(x).$$

2.2.2 *Details of a PUF Experiment*

A PUF response is generally considered a random variable. To assess the usability of a PUF class, information about the distribution of its PUF responses is needed. A possible way to obtain such knowledge is through experiment,[4] i.e. estimating distribution statistics from observed PUF response values.

---

[4] An alternative method of learning information about PUF response distributions is through physical modeling of the PUF class construction.

Definition and Parameters of a PUF Experiment

PUF response values observed in an experiment can be ordered in a number of different ways. Three important 'dimensions' in an array of observed PUF responses are discernable: (i) responses from different PUF instances; (ii) responses from the same PUF instance but on different challenges; and (iii) responses from the same PUF instance on the same challenge but from distinct evaluations.

Definition 1 (PUF Experiment) An $(N_{puf}, N_{chal}, N_{meas})$-experiment on a PUF class $\mathcal{P}$ is an array of PUF response values obtained through observation. An $(N_{puf}, N_{chal}, N_{meas})$-experiment contains $N_{puf} \times N_{chal} \times N_{meas}$ values, consisting of response evaluations on N_{puf} (random) PUF instances from the considered PUF class $\mathcal{P}$, challenged on the same set of N_{chal} (random) challenges, with N_{meas} distinct response evaluations for each challenge on each PUF instance.

$$\mathsf{Experiment}_{\mathcal{P}}(N_{puf}, N_{chal}, N_{meas}) \rightarrow \mathbf{Y}_{\mathsf{Exp}(\mathcal{P})} = \left[y_i^{(j)}(x_k) \leftarrow \mathsf{puf}_i(x_k).\mathsf{Eval}\left(r_j^{\mathsf{E}}\right) \right],$$

with

$$\forall 1 \leq i \leq N_{puf} : \mathsf{puf}_i \xleftarrow{\$} \mathcal{P},$$

$$\forall 1 \leq k \leq N_{chal} : x_k \xleftarrow{\$} \mathcal{X}_{\mathcal{P}},$$

$$\forall 1 \leq j \leq N_{meas} : r_j^{\mathsf{E}} \xleftarrow{\$} \{0, 1\}^*.$$

If the conditions of a PUF experiment are of importance, the condition specifier is mentioned as $\mathsf{Experiment}_{\mathcal{P}}^{\alpha}(N_{puf}, N_{chal}, N_{meas})$, which expresses that all PUF responses of the experiment are evaluated under these conditions.

2.2.3 PUF Response Intra-distance

Intra-distance Definition

Definition 2 A PUF response **intra**-distance is a random variable describing the distance between two PUF responses from the same PUF instance and using the same challenge:

$$D_{\mathsf{puf}_i}^{\mathsf{intra}}(x) \overset{\triangle}{=} \mathbf{dist}\left[Y_i(x); Y_i'(x) \right],$$

with $Y_i(x)$ and $Y_i'(x)$ two distinct and random evaluations of PUF instance puf_i on the same challenge x. Additionally, the PUF response intra-distance for a random PUF instance and a random challenge is defined as the random variable:

$$D_{\mathcal{P}}^{\mathsf{intra}} \overset{\triangle}{=} D_{\mathsf{PUF} \leftarrow \mathcal{P}}^{\mathsf{intra}}(X \leftarrow \mathcal{X}_{\mathcal{P}}).$$

dist[·; ·] can be any well-defined and appropriate distance metric over the response set $\mathcal{Y}$. In this book, responses are nearly always considered as bit vectors and the distance metric used is the Hamming distance or the fractional Hamming distance (cf. Sect. A.1.1).

Intra-distance Statistics

For the design of nearly all PUF-based applications, knowledge about the distribution of $D_{\mathcal{P}}^{\text{intra}}$ is of great importance for characterizing the *reproducibility* of a PUF class (cf. Sect. 3.2.2). Therefore, estimated descriptive statistics of this distribution are commonly presented as one of the most important basic quality metrics for a PUF class. Based on the observed responses $\mathbf{Y}_{\text{Exp}(\mathcal{P})}$ of a PUF experiment Experiment$_{\mathcal{P}}(N_{\text{puf}}, N_{\text{chal}}, N_{\text{meas}})$, a new array $\mathbf{D}_{\text{Exp}(\mathcal{P})}^{\text{intra}}$ of observed response intra-distances can be calculated:

$$\mathbf{D}_{\text{Exp}(\mathcal{P})}^{\text{intra}} = \left[\mathbf{dist}\left[y_i^{(j_1)}(x_k); y_i^{(j_2)}(x_k)\right]\right]_{\forall 1 \leq i \leq N_{\text{puf}}; \forall 1 \leq k \leq N_{\text{chal}}; \forall 1 \leq j_1 \neq j_2 \leq N_{\text{meas}}}.$$

From this array of observed response intra-distances, the following descriptive statistics are often calculated to provide an estimation of the underlying distribution parameters of $D_{\mathcal{P}}^{\text{intra}}$:

- An estimate of the distribution mean of $D_{\mathcal{P}}^{\text{intra}}$ or $\mathsf{E}[D_{\mathcal{P}}^{\text{intra}}]$ is obtained from the sample mean of $\mathbf{D}_{\text{Exp}(\mathcal{P})}^{\text{intra}}$:

$$\mu_{\mathcal{P}}^{\text{intra}} = \overline{\mathbf{D}_{\text{Exp}(\mathcal{P})}^{\text{intra}}} = \frac{2}{N_{\text{puf}} \cdot N_{\text{chal}} \cdot N_{\text{meas}} \cdot (N_{\text{meas}} - 1)} \cdot \sum \mathbf{D}_{\text{Exp}(\mathcal{P})}^{\text{intra}}.$$

- Equivalently, an estimate of the standard deviation of the distribution of $D_{\mathcal{P}}^{\text{intra}}$ or $\Sigma[D_{\mathcal{P}}^{\text{intra}}]$ is obtained as:

$$\sigma_{\mathcal{P}}^{\text{intra}} = \sqrt{\frac{2}{N_{\text{puf}} \cdot N_{\text{chal}} \cdot N_{\text{meas}} \cdot (N_{\text{meas}} - 1) - 2} \cdot \sum \left(\mathbf{D}_{\text{Exp}(\mathcal{P})}^{\text{intra}} - \mu_{\mathcal{P}}^{\text{intra}}\right)^2}.$$

- A estimate of the shape of the underlying distribution of $D_{\mathcal{P}}^{\text{intra}}$ is given by the histogram of $\mathbf{D}_{\text{Exp}(\mathcal{P})}^{\text{intra}}$.
- Order statistics of $\mathbf{D}_{\text{Exp}(\mathcal{P})}^{\text{intra}}$ give more robust information about the distribution of $D_{\mathcal{P}}^{\text{intra}}$ when it is skewed. Regularly used order statistics are the minimum, the maximum and the median, and for more detail the 1/4- and 3/4-quantiles and the 1 %- and 99 %-percentiles. In particular the maximum and the 99 %-percentile of $\mathbf{D}_{\text{Exp}(\mathcal{P})}^{\text{intra}}$ are of interest since they provide a good estimate of the largest PUF response intra-distances, i.e. the 'least reliable' PUF response one can expect.

Intra-distance Statistics Under Variable Evaluation Conditions

Variations in evaluation conditions such as environment temperature and supply voltage generally influence the intra-distance between PUF responses. The distance between two PUF responses evaluated on the same PUF instance and for the same challenge, but under different conditions α_1 and α_2, is typically larger than the intra-distance between the same responses evaluated under one fixed condition. For PUF-based applications, it is important to learn the worst-case, i.e. the largest, intra-distance which can arise for a given range of evaluation conditions with respect to a particular reference condition α_{ref}.

Definition 3 The PUF response intra-distance under condition α with respect to a reference condition α_{ref} is a random variable defined as:

$$D^{intra}_{puf_i;\alpha}(x) \triangleq \mathbf{dist}\big[Y_i^{\alpha_{ref}}(x); Y_i^{\alpha}(x)\big],$$

with $Y_i^{\alpha_{ref}}(x) \leftarrow puf_i(x).\mathsf{Eval}^{\alpha_{ref}}$ and $Y_i^{\alpha}(x) \leftarrow puf_i(x).\mathsf{Eval}^{\alpha}$ two distinct evaluations of PUF instance puf_i on the same challenge x but under different conditions α_{ref} and α. The PUF response intra-distance for a random PUF instance and a random challenge is defined as the random variable:

$$D^{intra}_{\mathcal{P};\alpha} \triangleq D^{intra}_{PUF\leftarrow\mathcal{P};\alpha}(X \leftarrow \mathcal{X}_{\mathcal{P}}).$$

All earlier introduced intra-distance statistics can be extended in the same manner to obtain estimates of the distribution of $D^{intra}_{\mathcal{P};\alpha}$. In practice, the nominal operating condition is selected as the reference condition. To find the worst-case intra-distance in a given range of conditions, one typically evaluates the intra-distance under the extrema of this range. This makes sense under the reasonable assumption that intra-distance with respect to the reference condition increases if one moves further away from the reference. The extrema of a range of conditions are called the *corner cases*. As an example, assume one needs to find the worst-case intra-distance behavior over the environment temperature range $T_{env} = -40\ °C\ldots85\ °C$. The reference condition is set at room temperature, $\alpha_{ref} = (T_{env} = 25\ °C)$, and the intra-distances at the corner cases $\alpha_1 = (T_{env} = -40\ °C)$ and $\alpha_2 = (T_{env} = 85\ °C)$ are studied.

2.2.4 PUF Response Inter-distance

Inter-distance Definition

Definition 4 A PUF response **inter**-distance is a random variable describing the distance between two PUF responses from different PUF instances using the same challenge:

$$D^{inter}_{\mathcal{P}}(x) \triangleq \mathbf{dist}\big[Y(x); Y'(x)\big],$$

with $Y(x)$ and $Y'(x)$ evaluations of the same challenge x on two random but distinct PUF instances $\mathsf{PUF} \leftarrow \mathcal{P}$ and $\mathsf{PUF}'(\neq \mathsf{PUF}) \leftarrow \mathcal{P}$. Additionally, the PUF response inter-distance for a random challenge is defined as the random variable:

$$D_{\mathcal{P}}^{\text{inter}} \triangleq D_{\mathcal{P}}^{\text{inter}}(X \leftarrow \mathcal{X}_{\mathcal{P}}).$$

Inter-distance Statistics

Again, the distribution of the random variable $D_{\mathcal{P}}^{\text{inter}}$ is an important metric for a PUF class, in this case to characterize the *uniqueness* of its instances. Based on the observed responses $\mathbf{Y}_{\text{Exp}(\mathcal{P})}$ of a PUF experiment $\mathsf{Experiment}_{\mathcal{P}}(N_{\text{puf}}, N_{\text{chal}}, N_{\text{meas}})$, a new array $\mathbf{D}_{\text{Exp}(\mathcal{P})}^{\text{inter}}$ of observed response inter-distances can be calculated:

$$\mathbf{D}_{\text{Exp}(\mathcal{P})}^{\text{inter}} = \left[\mathbf{dist}\left[y_{i_1}^{(j)}(x_k); y_{i_2}^{(j)}(x_k)\right]\right]_{\forall 1 \leq i_1 \neq i_2 \leq N_{\text{puf}}; \forall 1 \leq k \leq N_{\text{chal}}; \forall 1 \leq j \leq N_{\text{meas}}}.$$

From this sample array of response inter-distances, the following descriptive statistics can be calculated to provide an estimation of the underlying distribution parameters of $D_{\mathcal{P}}^{\text{inter}}$:

- The sample mean of $\mathbf{D}_{\text{Exp}(\mathcal{P})}^{\text{inter}}$ as an estimate of $\mathsf{E}[D_{\mathcal{P}}^{\text{inter}}]$:

$$\mu_{\mathcal{P}}^{\text{inter}} = \overline{\mathbf{D}_{\text{Exp}(\mathcal{P})}^{\text{inter}}} = \frac{2}{N_{\text{puf}} \cdot (N_{\text{puf}} - 1) \cdot N_{\text{chal}} \cdot N_{\text{meas}}} \cdot \sum \mathbf{D}_{\text{Exp}(\mathcal{P})}^{\text{inter}}.$$

- Equivalently, an estimate of $\mathbf{\Sigma}[D_{\mathcal{P}}^{\text{inter}}]$ is obtained as:

$$\sigma_{\mathcal{P}}^{\text{inter}} = \sqrt{\frac{2}{N_{\text{puf}} \cdot (N_{\text{puf}} - 1) \cdot N_{\text{chal}} \cdot N_{\text{meas}} - 2} \cdot \sum \left(\mathbf{D}_{\text{Exp}(\mathcal{P})}^{\text{inter}} - \mu_{\mathcal{P}}^{\text{inter}}\right)^2}.$$

- A estimate of the shape of the underlying distribution of $D_{\mathcal{P}}^{\text{inter}}$ is given by the histogram of $\mathbf{D}_{\text{Exp}(\mathcal{P})}^{\text{inter}}$.
- The order statistics of $\mathbf{D}_{\text{Exp}(\mathcal{P})}^{\text{inter}}$ again give more robust information about the distribution of $D_{\mathcal{P}}^{\text{inter}}$ when it is skewed. For PUF response inter-distances, in particular the minimum and the 1 %-percentile of $\mathbf{D}_{\text{Exp}(\mathcal{P})}^{\text{inter}}$ are of interest since they are good estimates of the smallest inter-distance, i.e. the 'least unique' PUF instance pair, one can expect.

Inter-distance Statistics Under Variable Evaluation Conditions

The inter-distance between PUF responses is also susceptible to variations in evaluation conditions. However, for typical PUF-based applications, the initial generation or enrollment is done in a secure environment under controlled conditions, and only the inter-distance of evaluations under the reference condition α_{ref} is of importance. Therefore, no extension of the definition of inter-distance to varying conditions is required in that case. In applications where the initial enrollment is done in the field, the susceptibility of the inter-distance to the evalution conditions does need to be taken into account. In this book, we only consider enrollment at reference conditions.

2.3 Terminology and Classification

In Sect. 2.3.1, we elaborate on the origin and the meaning of the acronym 'PUF'. As will become clear from this chapter, the growing popularity of PUFs has caused a proliferation of different constructions and concepts all labelled as PUFs. In an attempt to bring some order into this zoo, a number of classification attempts were made:

- Based on the implementation technology of the proposed construction: non-electronic PUF versus electronic PUFs versus silicon PUFs, as discussed in Sect. 2.3.2.
- Based on a more general set of physical construction properties: non-intrinsic versus intrinsic PUFs, as discussed in Sect. 2.3.3.
- Based on the algorithmic properties of their challenge-response behavior: weak versus strong PUFs, as discussed in Sect. 2.3.4.

2.3.1 "PUFs: Physical(ly) Unclon(e)able Functions"

To PUF or Not to PUF?

In addition to proposing a new optics-based PUF construction, Pappu [104] in 2001 was the first to also describe and define the more general concept of a PUF, which he introduced at the time as a *physical one-way function*. Shortly after, Gassend et al. [42] proposed a new silicon-based PUF construction and defined it similarly as a *physical random function*, but they opted for the acronym *PUF*, standing for *physical unclonable function*, to avoid confusion with the concept of a pseudo-random function which in cryptography is already abbreviated as PRF.

In the meantime, the acronym 'PUF' has become a catch-all for a variety of often very different constructions and concepts, but which share a number of interesting properties. Some of these constructions were proposed before the term PUF was coined, and were hence not called PUFs from the beginning. Other constructions were proposed in fields other than cryptographic hardware, where the term PUF is yet unknown. Depending on which properties are assumed to be defining for PUFs (and depending on whom you ask), such constructions are still called PUFs.

PUF having been used as a label for many different constructions, the literal semantic meaning of the acronym as "physical unclonable function" has been partially lost, up to the point where some consider it a misnomer. Moreover, slight variations in the actual wording were introduced over time, as expressed in the title of this section. In this book, the acronym PUF does not generally refer to its literal meaning. Instead, it is used as the collective noun for the variety of proposed constructions sharing a number of interesting properties. One of the main goals of the next chapter is to list known PUF constructions and identify a least common subset of such defining properties.

Physical Versus Physically?

Originally, the acronym PUF stood for *physical* unclonable function, but later the variant *physically* unclonable function also got into use. Both terms are still used today and refer to the same concept. The choice for one or the other depends mainly on the preference of the author. However, from a strictly semantic point of view, there is a small difference in meaning. A *physical* unclonable function is 'a physical function which is unclonable' whereas a *physically* unclonable function is 'a function which is physically unclonable'. In Sect. 3.2.10, we argue why we consider the second interpretation to be a more fitting description of the actual concept of a PUF.

Unclon(e)able?

The adjective *unclonable* is the central specifier in the acronym, reflecting the distinguishing property of a PUF. However, the actual meaning of unclonable is not clearly defined. In a broader sense, one first needs to define what one considers a *clone*. Especially in the context of PUFs, this has led to some debate. In many scenarios, one solely observes the input-output (or challenge-response) behavior of a PUF and only rarely the actual physical entity. In that respect, any construction which accurately mimics the behavior of a particular PUF could be considered a clone thereof. On the other hand, if one takes the physical aspect into account, cloning gets a different meaning, e.g. one could require a clone to have the same behavior *and* the same physical appearance as the original, or even more strictly, one could require a clone to be manufactured in the same production process as the original. The distinction between the former and latter interpretation of a clone is often made specific as a *mathematical* versus a *physical* clone.

Whether or not to write the silent '*e*' in unclon(e)able is somewhat arbitrary. Generally, the terminal, silent -*e* is dropped when appending the suffix -*able* to a verb [35], e.g. *movable*, *blamable*, unless the verb ends in -*ce* or -*ge*, which is not the case here. However, in scientific literature on computing the form (un)clon*eable* is more common, likely due to the existence of an homonymous but unrelated interface in the popular Java programming language [58]. A web search gives roughly a five-to-one preference of *unclonable* over *uncloneable*, although peculiarly for *clonable* versus *cloneable* these odds are reversed. In this book, the form *unclonable* is preferred.

Function?

In the pure mathematical sense, a PUF is not even strictly a *function*, since a single input (challenge) can be related to more than a single output (response) due to the uncontrollable effects of the physical environment and random noise on the response generation. Therefore, a more fitting mathematical description of a PUF is as a *probabilistic function*, i.e. a function for which part of the input is an uncontrollable random variable. This is sometimes made explicit by providing an additional input r to a function, representing an (undetermined) number of fair coin tosses: $r \xleftarrow{\$} \{0, 1\}^*$. Alternatively, one can consider the outcome of a probabilistic function to be a random variable with a particular probability distribution, instead of a deterministic value.

2.3.2 Non-electronic, Electronic and Silicon PUFs

PUFs and PUF-like constructions have been proposed based on a wide variety of technologies and materials such as glass, plastic, paper, electronic components and silicon integrated circuits (ICs). A possible classification of PUF constructions is based on the *electronic* nature of their identifying features.

Non-electronic PUFs

A first, relatively large class contains PUF proposals with an inherently non-electronic nature. Their PUF-like behavior is based on non-electronic technologies or materials, e.g. the random fiberstructure of a sheet of paper or the random reflection of scattering characteristics of an optical medium (cf. Sect. B.1). Note that the denomination non-electronic reflects the origin of the PUF behavior, and not the way PUF responses are processed or stored, which often does use electronics and digital circuitry.

Electronic PUFs

On the other side of the spectrum are electronic PUFs, i.e. constructions containing random variations in their electronic characteristics, e.g. resistance, capacitance, etc. These variations are generally also measured in an electronic way. Note that there is not always a sharp distinction between electronic and non-electronic PUFs, e.g. a construction called RF-DNA [29] consists of randomly arranged copper wires fixated in a flexible silicone medium which could be considered non-electronic, whereas LC-PUFs [46] are in fact passive LC resonance circuits which could be considered electronic in nature. However, the manner in which they generate responses is practically the same, i.e. the random influence they have on an imposed wireless radio-frequency field (cf. Sect. B.2). Another side note is that variation in electronic behavior is practically always a consequence of variations in non-electronic parameters at a lower level of physical abstraction, e.g. the length and section of a metal wire determine its resistance, the area and distance between two conductors determine their capacitance, etc.

Silicon PUFs

A major subclass of electronic PUFs are the so-called silicon PUFs, i.e. integrated electronic circuits exhibiting PUF behavior which are embedded on a silicon chip. The term silicon PUF and the first practical realization of a PUF structure in silicon were introduced by Gassend et al. [42]. Since silicon PUFs can be directly connected to standard digital circuitry embedded on the same chip, they can be immediately deployed as a hardware building block in cryptographic implementations. For this reason, it is evident that silicon PUFs are of particular interest for security solutions and they are the main type of PUF construction we focus on in this book.

2.3.3 Intrinsic and Non-intrinsic PUFs

An important classification of PUFs based on their construction properties is that of *intrinsic PUFs* as initially proposed by Guajardo et al. [45]. In a slightly adapted form, we consider that a PUF construction needs to meet at least two conditions to be called an intrinsic PUF: (i) its evaluations are performed *internally* by embedded measurement equipment, and (ii) its random instance-specific features are *implicitly* introduced during its production process. Next, we discuss both properties in more detail and highlight the specific practical and security advantages of intrinsic PUFs.

External Versus Internal Evaluations

A PUF evaluation is a physical measurement which can take many forms. We distinguish between external and internal evaluations. An *external* evaluation is a measurement of features which are externally observable and/or which is carried out using equipment which is external to the physical entity containing the features. On the other side are *internal* evaluations, i.e. measurements of internal features which are carried out by equipment completely embedded in the instance itself. This might seem a rather arbitrary distinction, but there are some important advantages to internal evaluations:

- The first is a purely practical advantage. Having the measurement equipment embedded in every PUF instance means that every PUF instance can evaluate itself without any external limitations or restrictions. Internal evaluations are typically also more accurate since there is less opportunity for outside influences and measurement errors.

- Secondly, there is a rather important security advantage to internal evaluations, since the PUF responses originate inside the embedding object. This means that, as long as an instance does not disclose a response to the outside world, it can be considered an internal secret. This is of particular interest when the embedding object is a digital IC, since in that case the PUF response can be used immediately as a secret input for an embedded implementation of a cryptographic algorithm.

A possible disadvantage of an internal evaluation is that one needs to trust the embedded measurement equipment, since it is impossible to externally verify whether the measurement takes place as expected, this as opposed to an external evaluation, for which it is often possible to actually observe the measurement in progress.

Explicit Versus Implicit Random Variations

A second construction-based distinction considers the source of the randomness of the evaluated features in a PUF. The manufacturing process of the PUF instances can contain an *explicit* randomization procedure with the sole purpose of introducing random features which will later be measured when the PUF is evaluated. Alternatively, the measured random features can also be an *implicit* side effect of the standard production flow of the embedding object, i.e. they arise naturally as an artifact of uncontrollable effects during the manufacturing process. Again, there are some subtle but interesting advantages to PUF constructions based on implicit random variations:

- From a practical viewpoint, implicit random variations generally come at no extra cost since they are already (unavoidably) present in the production process. Introducing an explicit randomization procedure on the other hand can be costly, particularly due to timing constraints in the manufacturing flow of high-volume products.
- Implicit random variability in manufacturing or so-called *process variations* which occur during a production process are generally undesirable since they have a negative impact on yield. This means that manufacturers already take elaborate measures to reduce process variations as much as possible. However, as it turns out, completely avoiding any random effect during manufacturing is technically impossible and process variations are always present. As a consequence, PUF constructions based on implicit process variations have the interesting security advantage that even the manufacturers, having all details and full control over their production process, cannot remove or control the random features at the core of the PUF's functionality.

Non-intrinsic PUF Constructions

A variety of PUF constructions have been proposed which we label *non-intrinsic* according to this classification, e.g.:

- The optical PUF as proposed by Pappu et al. [104, 105], which is based on the unique speckle pattern which arises when shining a laser on a transparent material with randomly solved scattering particles.
- The coating PUF as proposed by Tuyls et al. [147], which is based on the unique capacitance between two metal lines in a silicon chip whose top layer consists of a coating with randomly distributed dielectric particles.

The optical PUF is non-intrinsic since it is externally evaluated (by observing the speckle pattern) and its random features are explicitly introduced (by solving the scattering particles). The coating PUF is also non-intrinsic: although it can be internally evaluated on a silicon chip, its randomness is still explicitly generated during production (by solving the dielectric particles). Since our main focus is on intrinsic PUFs, we will not go into more detail on the construction of non-intrinsic PUFs in the main text of this book. For completeness, we have added a detailed description of many known non-intrinsic PUF constructions in Appendix B.

2.3.4 Weak and Strong PUFs

A last classification of PUFs we discuss here is explicitly based on the security properties of their challenge-response behavior. The distinction between *strong* and *weak* PUFs was first introduced by Guajardo et al. [45], and further refined by Rührmair et al. [116]. Basically, a PUF is called a strong PUF if, even after giving an adversary access to a PUF instance for a prolonged period of time, it is still possible to come up with a challenge to which with high probability the adversary does not know the response. Note that this implies at least that: (i) the considered PUF has a very large challenge set, since otherwise the adversary can simply query all challenges and no unknown challenges are left, and (ii) it is infeasible to built an accurate model of the PUF based on observed challenge-response pairs, or in other words the PUF is unpredictable. PUFs which do not meet the requirements for a strong PUF, and in particular PUFs with a small challenge set, are consequentially called weak PUFs. An extreme case is a PUF construction which has only a single challenge. Such a construction is also called a physically obfuscated key or POK (cf. Sect. 2.5.1). It is evident that the security notion of a strong PUF is higher than that of a weak PUF, hence enabling more applications. However constructing a practical (intrinsic) strong PUF with strong security guarantees turns out to be very difficult, up to the point where one can consider it an open problem whether this is actually possible.

2.4 Intrinsic PUF Constructions

All known intrinsic PUF constructions are silicon PUFs based on random process variations occurring during the manufacturing process of silicon chips. The known intrinsic silicon PUF constructions can be further classified based on their operating principles:

- A first class of constructions measures random variations on the delay of a digital circuit; the constructions are therefore called *delay-based* silicon PUFs.
- A second class of silicon PUF constructions use random parameter variations between matched silicon devices, also called device mismatch, in bistable memory elements. These constructions are labelled *memory-based* silicon PUFs.
- Finally, we classify a collection of intrinsic silicon PUF proposals consisting of mixed-signal circuits. Their basic operation is of an analog nature, which means an embedded analog measurement is quantized with an analog-to-digital conversion to produce a digital response representation.

2.4.1 Arbiter PUF

Basic Operation and Construction

The arbiter PUF, as proposed by Lee et al. [43, 75, 78], is a type of delay-based silicon PUF. The idea behind an arbiter PUF is to explicitly introduce a race condition between two digital paths on a silicon chip. Both paths end in an *arbiter circuit* which is able to resolve the race, i.e. it determines which of the two paths was faster and outputs a binary value accordingly. If both paths are designed or configured to have (nearly) identical nominal delays, the outcome of the race, and hence the output of the arbiter, cannot be unambiguously determined based on the design. Two scenarios are possible:

1. Even when designed or configured with identical nominal delays, the actual delay experienced by an edge which travels the two paths will not be exactly equal. Instead, there will be a random delay difference between the two paths, due to the effect of random silicon process variations on the delay parameters. This random difference will determine the race outcome and the arbiter output. Since the effect of silicon process variations is random per device, but static for a given device, the delay difference and by consequence the arbiter output will be device-specific. This is the basis for the PUF behavior of an Arbiter PUF.
2. Due to their stochastic nature, there is a non-negligible possibility that by chance both delays are nearly identical. In that case, when two edges are applied simultaneously on both paths they will reach the arbiter virtually at the same moment. This causes the arbiter circuit to go into a *metastable* state, i.e. the logic output of the arbiter circuit is temporarily undetermined (from an electronic point of view, the voltage of the arbiter output is somewhere in between the levels for a logic high and low). After a short but random time, the arbiter leaves its metastable state and outputs a random binary value which is independent of the outcome of the race. Although random, in this case the arbiter's output is not device-specific and not static. This phenomenon is the cause of unreliability (noise) of the responses of an arbiter PUF.

Lee et al. [43, 75, 78] propose implementing the two delay paths as chains of *switch blocks*. A switch block connects two input signals to its two outputs in one of

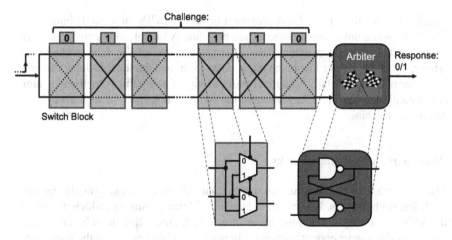

Fig. 2.1 Construction of a basic arbiter PUF as proposed by Lee et al. [43, 75, 78]

two possible configurations, straight or crossed, based on a configurable selection bit. The logic functionality of a switch block can be implemented in a number of ways, the most straightforward being with a pair of 2-to-1 multiplexers (muxes). By concatenating n switch blocks, a total of n configuration bits are needed to configure any of 2^n possible pairs of delay paths. This n-bit setting is considered the challenge of the arbiter PUF. Since the input-output delays of the two configurations of a switch block are slightly different and randomly affected by process variations, each of these 2^n challenges leads to a new race condition which can be resolved by the arbiter. The resulting output bit of the arbiter is considered the response of the arbiter PUF to the given challenge. However, it is immediately clear that these 2^n different configurations are not independent, since they are only based on a number of underlying delay parameters linear in n. This will lead to challenge-response modeling attacks on arbiter PUFs as we will discuss later. The arbiter circuit can also be implemented in a number of ways, the most straightforward being as an SR NAND latch. Most types of digital latches and flip-flops exhibit arbiter behavior. The construction of the basic arbiter PUF is depicted in Fig. 2.1

Ideally, an arbiter PUF response is solely determined by the random effect of process variations on the delay parameters of the circuit. This can only happen when two conditions are met:

1. The delay lines are designed to be nominally perfectly symmetrical, i.e. any difference in delay is solely caused by process variations.
2. The arbiter circuit is completely fair, i.e. it does not favor one of its inputs over the other. Lin et al. [80] conclude that a basic SR latch is the best option since it is unbiased due to its symmetric construction.

If either of the two conditions is not met, the arbiter PUF is biased, which results in a lower uniqueness of its responses. Designing a perfectly unbiased arbiter PUF is highly non-trivial since it requires very low-level control over implementation choices. In some technologies, e.g. on FPGAs [100], this is not possible, which greatly impedes the successful deployment of arbiter PUFs on them. If bias is unavoidable, some unbiasing techniques are possible. A tunable delay circuit in front of the arbiter PUF can counteract the deterministic bias of the delay paths or the arbiter, e.g. if the arbiter favors the upper delay path then the tunable delay can give the lower path a head start. A relatively simple way of doing this is by fixing a number of challenge bits in the beginning of the switch block chain in such a way as to counteract the bias.

Discussion on Modeling Attacks

The basic arbiter PUF construction with n chained switch blocks provides us with 2^n different challenges. However, the number of delay parameters which determine the arbiter PUF response is only linear in n, which means that these 2^n challenges cannot produce independent responses. In practice, when one learns the underlying delay parameters and one can model the interaction with the challenge bits, one is able to accurately predict response bits to random challenges, even without access to the PUF. In many ways, such a model can be considered a clone of the arbiter PUF, be it a mathematical clone. Arbiter PUF responses are only considered unpredictable as long as such a model cannot be accurately built.

It was immediately realized by Lee et al. [75] that their basic arbiter PUF can indeed be easily modeled and they even present modeling attack results for their own implementation. This and later modeling attacks assume a simple but accurate additive delay model for the arbiter PUF, i.e. the delay of a digital circuit is modeled as the sum of the delays of its serialized components. This turns out to be a very accurate first-order approximation for the relatively simple switch block chains. The unknown underlying delay parameters are learned from observing challenge-response pairs of the PUF. Every challenge-response pair constitutes a linear inequality over these parameters and is used to obtain a more accurate approximation. Moreover, the unknown parameters do not need to be calculated explicitly, but they are learned implicitly and automatically by an appropriate *machine-learning technique*. Machine-learning techniques such as artificial neural networks (ANNs) and support-vector machines (SVMs) are able to generalize behavior based on applied examples. Numerous results have been presented which successfully apply different machine-learning techniques in order to model the behavior of an arbiter PUF. The ultimate goal of such modeling attacks is to predict unknown responses as accurately as possible after having been trained with as few examples as possible. We say that a PUF is (p_{model}, q_{train})-modelable if a known modeling attack exists which is able to predict unseen responses of the PUF with a success rate p_{model}, after training the model with q_{train} known challenge-response pairs.

Experimental Results and Extended Constructions

Arbiter PUFs were initially implemented by Gassend et al. [43] on a set of 23 FPGA devices. From experimental results, the average intra- and inter-distance of the responses were estimated as $\mu_{\mathcal{P}}^{\text{intra}} = 0.1$ % and $\mu_{\mathcal{P}}^{\text{inter}} = 1.05$ %. The inter-distance is particularly low, with on average only one in 100 challenges producing a response bit which actually differs among different PUF instances. This arbiter PUF implementation is hence very biased, which is a result of the lack of low-level control over the placement and routing of the delay lines on the FPGA platform. A following implementation by the same team on a set of 37 ASICs proves to be less biased with $\mu_{\mathcal{P}}^{\text{inter}} = 23$ %, but is still far from the ideal of 50 %. The reliability of this implementation is very high, with $\mu_{\mathcal{P}}^{\text{intra}} = 0.7$ % under reference conditions. This worsens to $\mu_{\mathcal{P}}^{\text{intra}} = 4.82$ % when temperature rises to $T_{env} = 67$ °C. Lim [78] demonstrates that this arbiter PUF implementation is ($p_{\text{model}} = 96.45$ %, $q_{\text{train}} = 5000$)-modelable by applying an SVM-based machine-learning modeling attack. This result is problematic for this implementation, since the expected prediction error of 100 % − 96.45 % = 3.55 % is smaller than the average unreliability of the PUF under temperature variation. In practice, this means this implementation is not secure (after observing $q_{\text{train}} = 5000$ challenge-response pairs).

All subsequent work on arbiter PUFs is an attempt to make model-building attacks more difficult, by introducing non-linearities in the delays and by controlling and/or restricting the inputs and outputs to the PUF. Lee et al. [75] propose *feed-forward* arbiter PUFs as a first attempt to introduce non-linearities in the delay lines. They are an extension of their basic arbiter PUF, where the configuration of some switch blocks in the delay chains is not externally set by challenge bits, but is determined by the outcome of intermediate arbiters evaluating the race at intermediate points in the delay lines. This was equivalently tested on 37 ASICs, leading to $\mu_{\mathcal{P}}^{\text{inter}} = 38$ % and $\mu_{\mathcal{P}}^{\text{intra}} = 9.84$ % (at $T_{env} = 67$ °C). Note that the responses are much noisier, which is caused by the increased metastability since there are multiple arbiters involved. It was shown that the simple model-building attacks which succeeded in predicting the simple arbiter don't work any longer for this non-linear arbiter PUF. However, later results by Majzoobi et al. [93] and Rührmair et al. [118] show that with more advanced modeling techniques it is still possible to build an accurate model for the feed-forward arbiter PUF. Based on simulated implementations of feed-forward arbiter PUFs, Rührmair et al. [118] demonstrate them to be ($p_{\text{model}} > 97.5$ %, $q_{\text{train}} = 50000$)-modelable, with the actual success rate depending on the delay path length and the number of feed-forward arbiters. More advanced feed-forward architectures are proposed by Lao and Parhi [74].

Majzoobi et al. [94] present an elaborate discussion on techniques to make arbiter-based PUFs on FPGA more resistant to modeling. They use an initial device characterization step to choose the optimal parameters for a particular instantiation and use the reconfiguration possibilities of FPGAs to implement this. To increase randomness and to thwart model-building attacks, they use "hard-to-invert" input and output networks controlling the inputs to and outputs of the PUF, although these are not shown to be cryptographically secure. In particular, they consider multiple arbiter PUFs in parallel and take the exclusive-or (XOR) of the arbiter evaluations as the new response. By simulation, they show that this construction gives desirable PUF properties and makes model-building much harder. However,

Rührmair et al. [118] again show that model-building of these elaborate structures might be feasible by presenting a ($p_{model} = 99$ %, $q_{train} = 6000$)-model building attack based on a simulated implementation consisting of $n = 64$ switch blocks and three parallel arbiters which are XOR-ed together. For longer delay paths and more parallel arbiters, this modeling attack keeps on working but requires a considerable larger q_{train}.

Finally, a different approach towards model-building attacks for arbiter PUFs is taken by Öztürk et al. [48, 102, 103]. Instead of preventing the attack, they use the fact that a model of the PUF can be constructed as part of an authentication protocol. The proposed protocol is a variant of the protocol by Hopper and Blum [54], with security based on the difficulty of learning parity with noise. This is a highly unconventional use of a PUF as it explicitly assumes the PUF *can* be cloned mathematically, with the cloned PUF model acting as a noisy shared secret.

2.4.2 Ring Oscillator PUF

Basic Operation and Construction

A second type of delay-based intrinsic PUF is the ring oscillator PUF. A number of different constructions can be placed under this type, all based on measuring random variations on the frequencies of digital oscillating circuits. The source of this randomness is again the uncontrollable effect of silicon process variations on the delay of digital components. When these components constitute a ring oscillator, the frequency of this oscillation is equivalently affected. A typical ring oscillator PUF construction has two basic components, ring oscillators and frequency counters, which are arranged in a particular architecture and are often combined with a response-generating algorithm.

The first type of ring oscillator PUF was proposed by Gassend et al. [40, 42]. The ring oscillator they use is a variant of the switch block-based delay line as proposed for the arbiter PUF (cf. Sect. 2.4.1). This delay circuit is transformed into an oscillator by applying negative feedback. An additional AND-gate in the loop allows us to enable/disable the oscillation. The oscillating signal is fed to a frequency counter which counts the number of oscillating cycles in a fixed time interval. The resulting counter value is a direct measure of the loop's frequency. The construction of such a very basic ring oscillator PUF is shown in Fig. 2.2. Gassend et al. [40, 42] propose using a synchronously clocked counter. The oscillating signal is first processed by a simple edge detector which enables the counter every time a rising edge is detected. This architecture is robust, but the use of an edge detector limits the frequency of the ring oscillator to half the clock frequency.

The measured frequency of equally implemented ring oscillators on distinct devices shows sufficient variation due to process variations to act as a PUF response. However, it was immediately observed that the influence of environmental conditions on the evaluation of this construction is significant, i.e. changes in temperature

Fig. 2.2 Construction of a simple ring oscillator PUF as proposed by Gassend et al. [42]

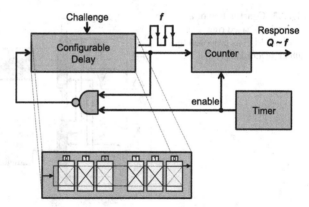

and voltage cause frequency changes which are orders of magnitude larger than those caused by process variations. To counter this influence, Gassend et al. [40, 42] propose an important post-processing technique called *compensated measuring*. The principle behind compensated measuring is to evaluate the frequency of two ring oscillators on the same device simultaneously and consider a differential function, i.c. the ratio, of both measurements as the eventual response. The reasoning behind this is that environmental changes will affect both frequencies in roughly the same way, and their ratio will be much more stable. Note that this type of differential measurement was already implied in the arbiter PUF construction considering two delay paths in parallel, which explains its relatively high reliability. For the proposed ring oscillator PUF construction, the compensated measurements based on the ratio of two frequencies also proves to be particularly effective. The average observed intra-distance (Euclidean) between evaluations is roughly two orders of magnitude smaller than the average inter-distance, even under varying temperature and voltage conditions.

Variants and Experimental Results

The results from Gassend et al. [42] clearly show the potential of ring oscillator PUFs. However, their proposed construction has some minor drawbacks. Firstly, since the ring oscillator is based on the same delay circuit as the simple arbiter PUF, this construction will also be susceptible to modeling attacks and similar counter-measures need to be included. Moreover, the produced response is an integer counter value, or a real value in the case of compensated measurement, and cannot be used directly as a bit string in subsequent building blocks. It first needs to be quantized in an appropriate way to obtain a bit string response with desirable distribution properties. Both these additions are indispensable in order to use this construction in an efficient and secure manner, but require additional resources.

Suh and Devadas [136] propose an alternative ring oscillator PUF architecture that is not susceptible to these modeling attacks and that naturally produces bitwise responses. Instead of a challengeable ring oscillator based on the delay circuit of

Fig. 2.3 Construction of a comparison-based ring oscillator PUF as proposed by Suh and Devadas [136]

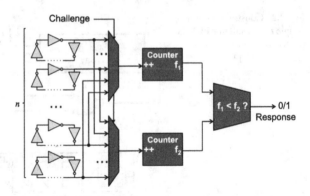

an arbiter PUF, their architecture contains an array of n fixed but identically implemented inverter chains. The architecture also contains two frequency counters, which can be fed by each of the n inverter chains. Two n-to-1 muxes control which oscillators are currently applied to the two counters. The selection signals of these muxes constitute the PUF's challenge. The schematic construction of this PUF is depicted in Fig. 2.3. As in the earlier proposal, both frequency counters are enabled for a fixed time interval and their resulting counter values are measures of the frequencies of the two applied oscillators. The PUF's response bit is generated by comparing the two counter values. Since the exact frequencies of the inverter chains are influenced by process variations, the resulting comparison bit will be random and device-specific. Note that the comparison operation is a very basic form of a compensated measurement, i.e. an inverter chain which is faster than another one under certain conditions is likely to be faster under all conditions.

By considering many pairwise combinations of inverter chains, this ring oscillator PUF construction can produce many response bits. If the PUF implements n ring oscillators, a total of $\binom{n}{2} = \frac{n \cdot (n-1)}{2}$ pairs can be formed. However, it is clear that all these pairs do not produce independent evaluations, e.g. if oscillator A is faster than B and B is faster than C, then it is apparent that A will also be faster than C and the resulting response bit is dependent on the previous two. Suh and Devadas [136] correctly state that the number of independent comparisons one can make is limited by the number of possible ways the frequencies of the oscillators can be ordered. If the frequencies are independent and identically distributed, each of $n!$ possible orderings is equally likely and the maximal number of independent comparisons one can make is $\log_2 n!$. However, the list of independent comparisons is difficult to obtain and is moreover device-specific, i.e. different oscillator pairs need to be compared on each instance of the PUF. A simpler and device-generic approach is to compare fixed pairs and use every oscillator only once. In this way, a ring oscillator PUF consisting of n oscillators will only produce $\frac{n}{2}$ response bits which are however guaranteed to be independent. Suh and Devadas [136] reduce the number of response bits even further by applying a post-processing technique called *1-out-of-k masking*, i.e. they evaluate k oscillators and only consider the pair with the largest difference in frequency and output the result of this comparison. This technique greatly enhances the reliability of the responses, but it comes at a relatively large resource overhead, i.e. out of n oscillators only $\lfloor \frac{n}{k} \rfloor$ response bits are generated. Experiments on a set of 15 FPGAs with $k = 8$ produce responses with an average inter-distance of 46.15 % and a very low average intra-distance of merely 0.48 % even under large temperature ($T_{env} = 120$ °C) and voltage ($V_{dd} + 10$ %) variations.

Maiti et al. [91] present results from a large-scale characterization of the ring oscillator PUF construction from Suh and Devadas [136] on FPGA. They do not apply the masking technique, but instead compare all neighbouring oscillators on the FPGA, yielding $n - 1$ response bits from n oscillators. Their results are particularly valuable as they are based on a large population of devices (125 FPGAs), which is rare in other works on PUFs. They present an average inter-distance of 47.31 % and an average intra-distance of 0.86 % at nominal conditions. However, changing temperature and especially voltage conditions have a significant impact on the intra-distance. At a 20 % reduced supply voltage, the average intra-distance goes up to 15 %. The measurement data resulting from this experiment is made publicly available [110].

Yin and Qu [157] build upon the construction from Suh and Devadas [136] and propose a number of adaptations which significantly improve the resource usage of the ring oscillator PUF. Firstly, they propose a generalization of the 1-out-of-k masking scheme which is much less aggressive in ignoring oscillators but achieves similar reliability results. The group of oscillators is partitioned into mutually exclusive subsets such that the minimal frequency difference between two members of a subset is above a certain threshold. An evaluation between a pair of oscillators from the same subset is therefore guaranteed to be very stable. Yet by using a clever grouping algorithm, practically all oscillators contribute to a response bit and only a few are ignored. Secondly, they propose a hybrid architecture for arranging ring oscillators and frequency counters. This allows us to find an optimal trade-off between the speed of generating responses and the resource usage of oscillators, counters and multiplexers.

Maiti and Schaumont [89, 90] also expand upon the construction from Suh and Devadas [136] and present some significant improvements. Firstly, they propose a number of guidelines for reducing systematic bias when implementing arrays of ring oscillators on an FPGA, and study the effect of each guideline on the uniqueness of the responses. Secondly, they propose a very efficient but highly effective variant of the 1-out-of-k masking scheme based on configurable ring oscillators. Instead of considering the most reliable pair out of k distinct oscillators, they make use of the oscillators' configurability and consider the most reliable out of k configurations of a single pair of oscillators. This achieves similar or even better reliability results than the original masking scheme, however without any sacrifice in resources.

A further improvement in the post-processing of ring oscillator PUFs is proposed by Maiti et al. [67, 92]. They go beyond ranking-based methods of oscillator frequencies by extracting a response based on the magnitude of the observed frequency differences. An elaborate post-processing algorithm based on test statistics of the observed frequency values and an identity mapping function is proposed. This substantially increases the amount of challenge-response pairs which can be considered for a limited set of ring oscillators. Maiti et al. [92] acknowledge that responses to different challenges are no longer information-theoretically independent, but they still show strong PUF behavior in terms of their inter- and intra-distance distributions. Experimental data obtained from an implementation on 125 FPGAs with 65519 challenge evaluations demonstrates an average inter-distance of 49.99 % and an average intra-distance of 10 % at $T_{env} = 70$ °C. The proposed post-processing technique is computationally intensive, requiring a sophisticated datapath architecture.

2.4.3 Glitch PUF

A third type of delay-based PUF construction is based on glitch behavior of combinatorial logic circuits. A purely combinatorial circuit has no internal state, which means that its steady-state output is entirely determined by its input signals. However, when the logical value of the input changes, transitional effects can occur, i.e. it can take some time before the output assumes its steady-state value. These effects are called glitches and the occurrence of glitches is determined by the differences in delay of the different logical paths from the inputs to an output signal. Since the exact circuit delays of a particular instance of a combinatorial circuit are influenced by random process variations, the occurrence, the number and the shape of the glitches on its output signals will equivalently be partially random and instance-specific. When accurately measured, the glitch behavior of such a circuit can be used as a PUF response.

Anderson [1] proposes a glitch-based PUF construction specifically for FPGA platforms. A custom logical circuit is implemented which, depending on the delay variations in the circuit, does or does not produce a single glitch on its output. The output is connected to the preset signal of a flip-flop which captures the glitch, should it occur. This is the circuit's single response bit. Placing many of these (small) circuits on an FPGA allows us to produce many PUF response bits. A challenge selects a particular circuit. Experimental results of 36 PUF implementations each producing 128 response bits show an average inter-distance of 48.3 % and an average intra-distance of 3.6 % under high temperature conditions ($T_{env} = 70$ °C).

Shimizu et al. [129, 137] present a more elaborate glitch PUF construction and also introduce the term *glitch PUF*. They propose a methodology to use the glitch behavior of any combinatorial circuit as a PUF response and apply it specifically to a combinatorial FPGA implementation of the SubBytes operation of AES [28]. Their initial proposal [137] consists of an elaborate architecture which performs an on-chip high-frequency sampling of the glitch wave form and a quantization circuit which generates a response bit based on the sampled data. They propose using the parity of the number of detected glitches as a random yet robust feature of a glitch wave form. In a later version of their PUF construction [129], they propose a much simpler implementation which achieves basically the same functionality and results. By simply connecting a combinatorial output to a toggle flip-flop, the value of the

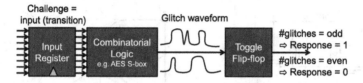

Fig. 2.4 Construction of a glitch PUF as proposed by Shimizu et al. [129]

toggle flip-flop after the transitional phase will equal the parity of the number of glitches that occurred. This construction is shown in Fig. 2.4. To improve the reliability of the PUF responses, a *bit-masking* technique is used. During an initial measurement on every instance, unstable response bits are identified as responses that do not produce a stable value on m consecutive evaluations. In later measurements, these response bits are ignored, i.e. they are *masked*. The mask information needs to be stored alongside each PUF instance or in an external storage. This type of masking also causes a resource overhead, as the identified unstable bits are not used in the PUF responses. However, the resulting unmasked bits show a greatly improved reliability. Experimental results from 16 FPGA implementations of this glitch PUF construction on 2048 challenges show an average inter-distance of 35 % and an average intra-distance of 1.3 % at nominal conditions, when bit-masking is applied. The bit-masking results in about 38 % of the response bits being ignored. Under variations of the environmental conditions, the average intra-distance increases substantially, to about 15 % in the worst-case corner condition of $T_{env} = 85\ °C$ and $V_{dd} + 5\ \%$.

2.4.4 SRAM PUF

SRAM Background

Static Random-Access Memory or SRAM is a digital memory technology based on bistable circuits. In a typical CMOS implementation, an individual SRAM cell is built with six transistors (MOSFETs), as shown in Fig. 2.5a. The logic memory functionality of a cell comes from two cross-coupled inverters at its core, shown in Fig. 2.5c, each built from two MOSFETs, one p-MOS and one n-MOS. From an electronic viewpoint, this circuit contains a positive feedback loop which reinforces its current state. In a logic sense, this circuit has two stable values (bistable), and by residing in one of the two states the cell stores one binary digit. Two additional access MOSFETs are used to read and write its contents. Typically, many SRAM cells are arranged in large memory array structures, capable of storing many kilobits or megabits. An SRAM cell is volatile, meaning that its state is lost shortly after power-down.

The detailed operation of an SRAM cell is best explained by drawing the voltage transfer curves of the two cross-coupled inverters, shown in Fig. 2.5b. From this graph, it is clear that the cross-coupled CMOS inverter structure has three possible operating points of which only two are stable and one is metastable. The sta-

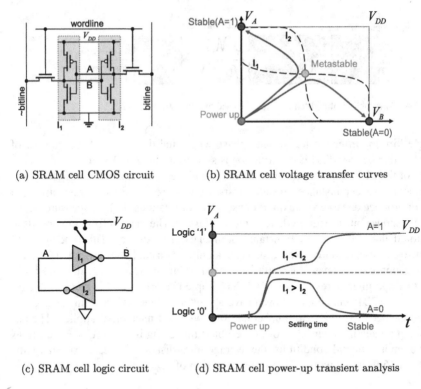

(a) SRAM cell CMOS circuit   (b) SRAM cell voltage transfer curves

(c) SRAM cell logic circuit   (d) SRAM cell power-up transient analysis

Fig. 2.5 Construction and power-up behavior of an SRAM cell

ble points are characterized by the property that deviations from these points are
reduced and the stable point condition is restored. This does not hold for the
metastable point, i.e. any small deviation from the metastable point is immediately
amplified by the positive feedback and the circuit moves away from the metastable
point towards one of the two stable points. Since electronic circuits are constantly
affected by small deviations due to random noise, an SRAM cell will never stay in
its metastable state very long but will quickly end up in one of the two stable states
(randomly).

Basic Operation

The operation principle of an SRAM PUF is based on the transient behavior of an
SRAM cell when it is powered up, i.e. when its supply voltage V_{DD} comes up. The
circuit will evolve to one of its operating points, but it is not immediately clear to
which one. The preferred initial operating point of an SRAM cell is determined by
the difference in 'strength' of the MOSFETs in the cross-coupled inverter circuit.
The transient behavior of an SRAM cell at power-up is shown in Fig. 2.5d. For efficiency and performance reasons, typical SRAM cells are designed to have perfectly

matched inverters. The actual difference in strength between the two inverters, the so-called *device mismatch*, is caused by random process variations in the silicon production process and is hence cell-specific. Each cell will have a random preferred initial operating point.

When one of the inverters is significantly stronger than the other one, the preferred initial operating point will be a stable state and the preference will be very distinct, i.e. such a cell will always power-up in the same stable state, but which state this is ('0' or '1') is randomly determined for every cell. When the mismatch in a cell is small, the effect of random circuit noise comes into play. Cells with a small mismatch still have a preferred initial stable state which is determined by the sign of the mismatch, but due to voltage noise there is a non-negligible probability that they power-up in their non-preferred state. Finally, cells which, by chance, have a negligible mismatch between their inverters, will power up in, or very close to, the metastable operating point. Their final stable state will be largely random for every power-up.

The magnitude of the impact of process variations on random device mismatch in SRAM cells causes most cells to have a strongly preferred but cell-specific initial state, and only few cells have a weak preference or no preference at all. This means that the power-up state of a typical SRAM cell shows strong PUF behavior. Large arrays of SRAM cells are able to provide thousands to millions of response bits for this SRAM PUF. The address of a specific cell in such an array can be considered the challenge of the SRAM PUF.

Results

SRAM PUFs were proposed by Guajardo et al. [45] and a very similar concept was simultaneously presented by Holcomb et al. [52]. Guajardo et al. [45] collect the power-up state of 8190 bytes of SRAM from different memory blocks on different FPGAs. The results show an average inter-distance between two different blocks of 49.97 %, and the average intra-distance within multiple measurements of a single block is 3.57 % at nominal conditions and at most 12 % for large temperature deviations. Holcomb et al. [52, 53] study the SRAM power-up behavior on two different platforms: a commercial off-the-shelf SRAM chip and embedded SRAM in a microcontroller chip. For 5120 blocks of 64 SRAM cells measured on eight commercial SRAM chips at nominal conditions, they obtained an average inter-distance of 43.16 % and an average intra-distance of 3.8 %. For 15 blocks of 64 SRAM cells from the embedded memory in three microcontroller chips, they obtained $\mu_{\mathcal{P}}^{\text{inter}} = 49.34$ % and $\mu_{\mathcal{P}}^{\text{intra}} = 6.5$ %.

SRAM PUFs were tested more extensively by Selimis et al. [126] on 68 SRAM devices implemented in 90 nm CMOS technology, including advanced reliability tests considering the ramp time of the supply voltage (t_{ramp}) and accelerated ageing of the circuit for an equivalent ageing time (t_{age}) of multiple years. Their test results show an average inter-distance $\mu_{\mathcal{P}}^{\text{inter}} \approx 50$ % and intra-distances of $\mu_{\mathcal{P}}^{\text{intra}} < 4$ % (nominal), $\mu_{\mathcal{P}}^{\text{intra}} < 19$ % ($T_{env} = -40$ °C), $\mu_{\mathcal{P}}^{\text{intra}} \approx 6$ % ($V_{dd} \pm 10$ %), $\mu_{\mathcal{P}}^{\text{intra}} < 10$ %

$(t_{ramp} = 1$ ms$)$ and $\mu_{\mathcal{P}}^{intra} < 14\ \%$ $(t_{age} = 4.7$ years$)$ respectively. Schrijen and van der Leest [124] also perform a large investigation into the reliability and uniqueness of SRAM PUFs across implementations in five different CMOS technologies (from 180 nm to 65 nm) coming from different SRAM vendors. For results we refer to their paper. Both these extensive studies prove the generally strong PUF behavior of SRAM power-up values over widely varying conditions and technologies.

2.4.5 Latch, Flip-Flop, Butterfly, Buskeeper PUFs

Besides SRAM cells, there are a number of alternative, more advanced digital storage elements which are based on the bistability principle. They all have the potential to display PUF behavior based on random mismatch between nominally matched cross-coupled devices.

Latch PUFs

Su et al. [135] present an IC identification technique based on the settling state of two cross-coupled NOR-gates which constitute a simple SR latch. By asserting a reset signal, this latch is forced into an unstable state and when released it will converge to a stable state depending on the internal mismatch between the NOR gates. Experiments on 128 NOR-latches implemented on 19 ASICs manufactured in 130 nm CMOS technology yield $\mu_{\mathcal{P}}^{inter} = 50.55\ \%$ and $\mu_{\mathcal{P}}^{intra} = 3.04\ \%$. An equivalent latch PUF cell structure based on cross-coupled NAND gates is possible, as shown in Fig. 2.6a.

A practical advantage of this latch PUF construction over SRAM PUFs is that the PUF behavior does not rely on a power-up condition, but can be (re)invoked at any time when the device is powered. This implies that secrets generated by the PUF need not be stored permanently over the active time of the device but can be recreated at any time. Moreover, it allows us to measure multiple evaluations of each response, which makes it possible to improve reliability through post-processing techniques such as majority voting. An interesting approach is taken by Yamamoto et al. [156], who use multiple evaluations of a latch PUF on FPGA to detect unreliable cells. Instead of discarding these cells, they consider them as a third possible power-up state, besides stable at '0' and stable at '1', which effectively increases the response length of the PUF by a factor of at most $\log_2(3)$.

Flip-Flop PUFs

In [84], we propose a PUF based on the power-up behavior of clocked D flip-flops on an FPGA platform. Most D flip-flop implementations are composed of a number of latch structures which are used to store a binary state, as shown in Fig. 2.6b. These internal latch structures cause the PUF behavior of a D flip-flop as for the basic latch

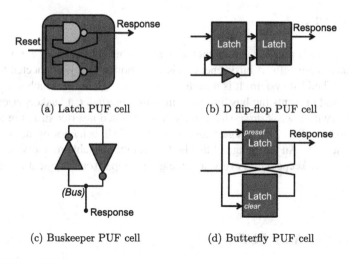

(a) Latch PUF cell   (b) D flip-flop PUF cell

(c) Buskeeper PUF cell   (d) Butterfly PUF cell

Fig. 2.6 Different PUFs based on bistable memory elements

PUF. We measure the power-up values of 4096 D flip-flops from three FPGAs and apply simple majority voting post-processing techniques, generating one response bit from 5 measurements on 9 flip-flops, to improve the uniqueness of the responses. This yields $\mu_{\mathcal{P}}^{\text{inter}} \approx 50$ % and $\mu_{\mathcal{P}}^{\text{intra}} < 5$ %.

van der Leest et al. [150] perform a more elaborate experimental study, measuring the power-up values of 1024 flip-flops on each of 40 ASIC devices and under varying conditions. Their experiments yield an average inter-distance of 36 % on the raw flip-flop values, and an average intra-distance strictly smaller than 13 % even under large temperature variations. They also propose a number of basic post-processing techniques to increase the uniqueness of the responses, including von Neumann extraction [153] and XOR-ing response bits.

Butterfly PUFs

For FPGA platforms, SRAM PUFs are often impossible because the SRAM arrays on most commercial FPGAs (when present) are forcibly cleared immediately after power-up. This results in the loss of any PUF behavior. In an attempt to provide an SRAM PUF-like construction which works on standard FPGAs, we propose the butterfly PUF [72]. The behavior of an SRAM cell is mimicked in the FPGA reconfigurable logic by cross-coupling two transparent data latches, forming a bistable circuit depicted in Fig. 2.6d. Using the preset/clear functionality of the latches, this circuit can be forced into an unstable state and will again converge when released. Measurement results on 64 butterfly PUF cells on 36 FPGAs yield $\mu_{\mathcal{P}}^{\text{inter}} \approx 50$ % and $\mu_{\mathcal{P}}^{\text{intra}} < 5$ % under high temperature conditions. It must be noted that, due to the discrete routing options on FPGAs, it is not trivial to implement the cell in such a way that the *mismatch by design* is small. This is a necessary condition if one wants the random mismatch caused by manufacturing variability to have any effect.

Buskeeper PUFs

Simons et al. [132] propose yet another variant of a bistable memory element PUF based on buskeeper cells. A buskeeper cell is a component that is connected to a bus line on an embedded system. It is used to maintain the last value which was driven on the line and prevents the bus line from floating. Internally, a buskeeper cell is a simple latch with a weak drive-strength. A cross-coupled inverter structure at their core, as shown in Fig. 2.6c, is again the cause of the PUF behavior of the power-up state of these cells. An advantage of this PUF over other bistable memory cell PUFs is that a basic buskeeper cell is very small, e.g. in comparison to typical latches and D flip-flops.

2.4.6 Bistable Ring PUF

The bistable ring PUF as proposed by Chen et al. [22] presents a combination of elements from both ring oscillator PUFs and SRAM PUFs. Its structure is very similar to that of a ring oscillator PUF, consisting of a challengeable loop of inverting elements. However, in contrast to the ring oscillator PUF, the number of inverting elements is even, which implies that the loop does not oscillate but exhibits bistability, like the SRAM PUF. Using reset logic, the bistable ring can be destabilized and after a settling time it stabilizes to one of the two stable states. The preferred stable state is again a result of process variations and hence instance-specific and usable as a PUF response. A single loop structure can be configured in many different ways, each exhibiting its own preferred stable state, and the exact configuration is controlled by the PUF's challenge. Chen et al. [22] observe that the settling time to reach a stable state is very challenge-dependent. Evaluating a response too quickly, before the ring is completely settled, results in severely reduced uniqueness. An implementation of a 64-element bistable ring PUF was configured on eight FPGAs and evaluated 12 times for each of 50000 different challenges. With a settling time of 47 μs, this results in an average observed intra-distance of 2.19 % at nominal conditions and 5.81 % at $T_{env} = 85$ °C, and an average inter-distance of 41.91 %. Due to the structural similarities between the bistable ring and a challengeable delay chain, as for the arbiter PUF, modeling attacks could be possible. Currently, no modeling results are known, but further analysis is required before strong claims regarding the unpredictability of the bistable ring PUF can be made.

2.4.7 Mixed-Signal PUF Constructions

Next, we describe a number of integrated electronic PUF proposals whose PUF behavior is inherently of an analog nature. This means that they also require embedded analog measurement techniques, and an analog-to-digital conversion if their responses are required in digital form, making them mixed-signal electronic constructions. While they are technically labelled as intrinsic PUFs, since they can be integrated on a silicon die and take their randomness from process variations, it is not straightforward to deploy these constructions in a standard digital silicon circuit, due to their mixed-signal nature and possibly due to non-standard manufacturing processes.

ICID: A Threshold Voltage PUF

To the best of our knowledge, Lofstrom et al. [82] were the first to propose an embedded technique called ICID which measures implicit process variations on integrated circuits and uses them as a unique identifier of the embedding silicon die. Their construction consists of a number of equally designed transistors laid out in an addressable array. The addressed transistor drives a resistive load and because of the effect of random process variations on the threshold voltages of these transistors, the current through this load will be partially random. The voltage over the load is measured and converted to a bit string with an auto-zeroing comparator. The technique was experimentally verified on 55 ASICs produced in 350 nm CMOS technology. An average intra-distance under extreme environmental variations of $\mu_{\mathcal{P}}^{intra} = 1.3\ \%$ is observed, while $\mu_{\mathcal{P}}^{inter}$ is very close to 50 %.

Inverter Gain PUF

Puntin et al. [106] present a PUF construction based on the variable gain of equally designed inverters, caused by process variations. The difference in gain between a pair of inverters is determined in an analog electronic measurement and converted into a single bit response. The proposed construction is implemented in 90 nm CMOS technology and an experimentally observed average intra-distance <0.07 % at nominal conditions and <0.4 % at ($T_{env} = 125\ °C$, $V_{dd} + 10\ \%$) is reported; however, this already includes a masking of the 20 % least stable response bits. The inter-distance is not presented, but it is mentioned that the correlation between response strings from different PUF instances is less than 1 %.

SHIC PUFs

Rührmair et al. [117, 121] propose a circuit consisting of an addressable array of diodes which is implemented as a crossbar memory in the aluminium-induced layer exchange (ALILE) technology. They observe that the read-out time of each memory cell, consisting of a single diode, is influenced by process variations and hence usable as a PUF response. Since each crosspoint in a dense grid of metal lines creates a diode, a very high physical density of response bits of up to 10^{10} bit/cm^2 can be obtained; hence the name Super-High Information Content or SHIC PUF. The performance of the construction is very low, with read-out speeds limited to merely 100 bit/s. The authors claim that the combination of high density and low performance provides protection against full read-out attacks, but it is unclear whether the particularly low performance makes any practical application possible. The physical mechanisms to produce such a SHIC PUF were tested, but no experimental PUF data is provided.

SRAM Failure PUF

An SRAM PUF variant is proposed by Fujiwara et al. [38]. The *static noise margin* (SNM) of an SRAM cell is a key figure of merit describing the resistance of a cell to voltage noise. Voltage noise exceeding the SNM alters a cell's state and hence deletes the information it is storing, resulting in a bit failure. The exact value of a cell's SNM is also influenced by silicon process variations; hence each cell in an SRAM array will produce a bit failure at a slightly different noise level. Fujiwara et al. [38] apply this observation to build a PUF. By gradually increasing the word line voltage of an SRAM array, the SNM of a cell is reduced until it produces a bit failure. They do this for a large array of cells and identify the first n failing cells. The addresses of these cells are random and are used as the PUF's responses. They present experimental results from 53 test ASICs, each producing 128 response bits, and report an average inter-distance of 49.92 %. The average observed intra-distance is as low as 0.1 %, but this is already after an elaborate majority voting post-processing. A very similar concept is proposed by Krishna et al. [70] as the MECCA PUF. They shorten the word line duty cycle of the SRAM array to induce failures.

2.4.8 Overview of Experimental Results

Table 2.1 summarizes all experimental results on intrinsic PUF constructions that are discussed in this section. We compare the average response intra- and inter-distances of the proposals since they are the most important quality parameters of a PUF and are therefore nearly always mentioned for a PUF experiment. When possible, we also mention the details of the experiment which produced these statistics, i.e. the condition under which the experiment was performed (α), and the size of the experiment in terms of number of different PUF instances (N_{puf}), number of challenges measured per instance (N_{chal}), and number of evaluations of each challenge (N_{meas}).

Although the intra- and inter-distance means give a good first notion of the quality of a PUF, there are a number of other measures which need to be considered when objectively comparing different PUF proposals. For one, the PUF proposals listed in Table 2.1 represent a wide variety of different implementations. Besides the basic PUF quality, the implementation efficiency and performance are equally important

Table 2.1 Overview of experimental results of intrinsic PUF constructions in the literature

Intrinsic PUF proposal $\mathcal{P}$	Experiment details				Experiment results	
	Condition α	N_{puf}	N_{chal}	N_{meas}	$\mu_{\mathcal{P}}^{inter}$	$\mu_{\mathcal{P};\alpha}^{intra}$
Simple Arbiter PUF [43]	nominal	23 (FPGA)	100000	200	0.10 %	1.05 %
Simple Arbiter PUF [43]	$T_{env} = 60\,°C$	23 (FPGA)	100000	23	0.30 %	1.05 %
Simple Arbiter PUF [75]	nominal	37 (ASIC)	10000	?	0.70 %	23.00 %
Simple Arbiter PUF [75]	$T_{env} = 67\,°C$	37 (ASIC)	10000	?	4.82 %	23.00 %
Simple Arbiter PUF [75]	$V_{dd} - 2\,\%$	37 (ASIC)	10000	?	3.74 %	23.00 %
FF Arbiter PUF [79]	nominal	20 (ASIC)	100000	?	4.50 %	38.00 %
FF Arbiter PUF [79]	$T_{env} = 67\,°C$	20 (ASIC)	100000	?	9.84 %	38.00 %
Basic ROPUF [42]	$T_{env} = 50\,°C$	4 (FPGA)	?	?	0.01 %	1.00 %
ROPUF [136]	$T_{env} = 120\,°C\,\&\,V_{dd} + 10\,\%$	15 (FPGA)	128	?	0.48 %	46.15 %
ROPUF [91]	nominal	125 (FPGA)	511	100	0.86 %	47.13 %
ROPUF [91]	$T_{env} = 65\,°C$	5 (FPGA)	511	100	4.00 %	47.13 %
ROPUF [91]	$V_{dd} - 20\,\%$	5 (FPGA)	511	100	15.00 %	47.13 %
Improved ROPUF [90]	nominal	5 (FPGA)	255	?	0.00 %	44.10 %
Improved ROPUF [90]	$T_{env} = 65\,°C$	5 (FPGA)	255	?	0.00 %	44.10 %
Improved ROPUF [90]	$V_{dd} - 20\,\%$	5 (FPGA)	255	?	2.00 %	44.10 %
Enhanced ROPUF [92]	$T_{env} = 70\,°C$	125 (FPGA)	65519	?	49.99 %	10 %
Glitch PUF [1]	$T_{env} = 70\,°C$	36 (FPGA)	128	?	3.60 %	48.30 %
Glitch PUF [129]	nominal	16 (FPGA)	2048	?	1.30 %	35.00 %
Glitch PUF [129]	$T_{env} = 85\,°C\,\&\,V_{dd} + 5\,\%$	16 (FPGA)	2048	?	15.00 %	35.00 %

Table 2.1 (*Continued*)

Intrinsic PUF proposal $\mathcal{P}$	Experiment details				Experiment results	
	Condition α	N_{puf}	N_{chal}	N_{meas}	$\mu_{\mathcal{P}}^{inter}$	$\mu_{\mathcal{P};\alpha}^{intra}$
SRAM PUF [45]	nominal	? (FPGA)	8190	92	3.57 %	49.97 %
SRAM PUF [45]	$T_{env} = 80\,°C$	? (FPGA)	8190	?	12.00 %	49.97 %
SRAM PUF [53]	nominal	5120 (COTS)	64	?	3.80 %	43.16 %
SRAM PUF [53]	nominal	15 (μC)	64	?	6.50 %	49.34 %
SRAM PUF [126]	nominal	68 (ASIC)	2048	?	<4 %	≈50.00 %
SRAM PUF [126]	$T_{env} = -40\,°C$	68 (ASIC)	2048	?	<19 %	≈50.00 %
SRAM PUF [126]	$V_{dd} \pm 10\,\%$	68 (ASIC)	2048	?	6.00 %	≈50.00 %
SRAM PUF [126]	$t_{ramp} = 1$ ms	68 (ASIC)	2048	?	<10 %	≈50.00 %
SRAM PUF [126]	$t_{age} = 14$ years	68 (ASIC)	2048	?	<14 %	≈50.00 %
Latch PUF [135]	nominal	19 (ASIC)	128	?	3.04 %	50.55 %
Flip-Flop PUF [84]	nominal	3 (FPGA)	4096	?	<5 %	≈50.00 %
Flip-Flop PUF [150]	$T_{env} = 80\,°C$	40 (ASIC)	1024	?	<13 %	36.00 %
Butterfly PUF [72]	$T_{env} = 80\,°C$	36 (FPGA)	64	?	<5 %	≈50.00 %
Bistable Ring PUF [22]	nominal	8 (FPGA)	50000	12	2.19 %	41.91 %
Bistable Ring PUF [22]	$T_{env} = 85\,°C$	8 (FPGA)	50000	12	5.81 %	41.91 %
ICID [82]	nominal	55 (ASIC)	132	?	1.30 %	≈50.00 %
SRAM Failure PUF [38]	nominal	53 (ASIC)	128	?	0.01 %	49.92 %

for most applications. However, for many PUF constructions, detailed implementation parameters are not provided, or implementation efficiency is even sacrificed to improve the PUF quality. This gives a distorted picture when comparing different PUFs solely based on Table 2.1. Moreover, the listed PUF constructions are also implemented on a variety of platforms and technologies. We differentiate the used platforms between field-programmable gate arrays (FPGAs), application-specific integrated circuits (ASICs), commercial off-the-shelf products (COTS) and microcontrollers (μCs). However, even within a single platform, different technologies can be used, e.g. the different scaling CMOS technology nodes. Finally, as is clear from the experiment details, the size of the experiments and hence the statistical significance of the obtained results also varies greatly. To overcome these issues and provide an objective comparison between intrinsic PUFs, we present an extensive experimental analysis of a number of intrinsic PUF constructions in Chap. 4.

2.5 PUF Extensions

2.5.1 POKs: Physically Obfuscated Keys

The concept of a physically obfuscated key or POK was introduced by Gassend [40], and generalized to physically obfuscated algorithms by Bringer et al. [14]. The only condition for a POK is that a key is permanently stored in a 'physical' way instead of a digital way, which makes it hard for an adversary to learn the key by a probing attack. Additionally, an invasive attack on the device storing the key should destroy the key and make further use impossible, hence providing tamper evidence. It is clear that POKs and PUFs are very similar concepts, and it has already been pointed out by Gassend [40] that POKs can be built from (tamper-evident) PUFs and vice versa.

2.5.2 CPUFs: Controlled PUFs

A controlled PUF or CPUF, as introduced by Gassend et al. [41], is in fact a mode of operation for a PUF in combination with other (cryptographic) primitives. A PUF is said to be controlled if it can only be accessed via an algorithm which is physically bound to the PUF in an inseparable way. Attempting to break the link between the PUF and the access algorithm should preferably lead to the destruction of the PUF. There are a number of advantages by turning a PUF into a CPUF:

- A (cryptographic) hash function to generate the challenges of the PUF can prevent chosen-challenge attacks, e.g. to make model-building attacks more difficult. However, for arbiter PUFs it has been shown that model-building attacks work equally well for randomly picked challenges.

- An error-correction algorithm acting on the PUF measurements makes the final responses much more reliable, reducing the probability of a bit error in the response to virtually zero.
- A (cryptographic) hash function applied on the error-corrected outputs effectively breaks the link between the responses and the physical details of the PUF measurement. This makes model-building attacks much more difficult. When hashing a PUF's response, error-correction is indispensable since any minor deviation on the response gives an entirely unrelated hash result.
- The hash function generating the PUF challenges can take additional inputs, e.g. allowing to give a PUF multiple personalities. This might be desirable when the PUF is used in privacy-sensitive applications.

2.5.3 RPUFs: Reconfigurable PUFs

Reconfigurable PUFs or RPUFs were introduced by Kursawe et al. [73]. The idea behind an RPUF is to extend the regular challenge-response behavior of a PUF with an additional operation called *reconfiguration*. This reconfiguration has the effect that the partial or complete challenge-response behavior of the PUF is randomly and preferably irreversibly changed, leading to a new PUF. Kursawe et al. [73] propose two possible implementations of RPUFs where the reconfiguration mechanism is an actual physical reconfiguration of the randomness in the PUF. One is an extension of optical PUFs, where a strong laser beam briefly melts the optical medium, causing a random rearrangement of the optical scatterers, which leads to a completely new optical challenge-response behavior. The second proposal is based on a type of non-volatile storage called phase change memory. Writing to such a memory amounts to physically altering the phase of a small cell from crystalline to amorphous or somewhere in between, and it is read out by measuring the resistance of the cell. Since the resistance measurements are more accurate than the writing precision, the exact measured resistances can be used as responses, and rewriting the cells will change them in a random way. Both proposals are rather exotic at this moment and remain largely untested. A third proposed option is actually a logical extension of a regular PUF. By fixing a part of a PUF's challenge with a fuse register, the PUF can be reconfigured by blowing a fuse, which optimally leads to a completely changed challenge-response behavior for the remaining external challenge bits.

Katzenbeisser et al. [64] generalize this last option in a logically reconfigurable PUF (LRPUF). An LRPUF consists of a regular PUF, a portion of non-volatile memory that stores a state and state-dependent input- and output-transformations respectively on the PUF's challenge and response. By updating the state of an LRPUF using an appropriate state-update mechanism, the input- and output-transformations change, resulting in a completely different challenge-response behavior for the whole construction. This state update is hence a logical form of reconfiguring the PUF. Katzenbeisser et al. [64] propose a number of practical input and output transformations and state update mechanisms which achieve interesting security properties. LRPUFs have some drawbacks with respect to actual *physically reconfigurable*

PUFs as proposed by Kursawe et al. [73]: (i) it is debatable whether the reconfiguration of an LRPUF is truly irreversible, since if an old internal state is restored, the same challenge-response behavior will reappear, and (ii) the security of the logical reconfiguration relies on the integrity of the state memory, which needs to be guaranteed by independent physical protection measures. However, their relatively easy construction in comparison to the rather exotic physically reconfigurable PUF proposals makes them immediately deployable.

A more specific form of a reconfigurable PUF is introduced by Rührmair et al. [120] as an *erasable PUF*. An erasable PUF is best described as a reconfigurable PUF with a reconfiguration granularity of a single challenge-response pair, i.e. the behavior of a single challenge-response pair of an erasable PUF instance can be (irreversibly) altered while keeping the behavior of all other pairs fixed. As demonstrated by Rührmair et al. [120], this construction enables some interesting security features for protocols based on erasable PUFs.

2.5.4 PPUFs: Public PUFs and SIMPL Systems

A number of attempts to use PUFs as part of a public-key-like algorithm have been made. SIMPL systems were proposed by Rührmair [112] and are an acronym for *Simulation Possible but Laborious*. Two potential implementations of such a system are discussed by Rührmair et al. [114]. A very similar concept was proposed by Beckmann and Potkonjak [6] as Public PUFs or PPUFs. Both SIMPL systems and PPUFs rely on physical challenge-response systems (PPUFs) which can be modeled explicitly, but for which evaluating the model is laborious and takes a detectably longer amount of time than the evaluation of the PPUF itself.

2.6 Conclusion

2.6.1 Overview of PUF Constructions

In this chapter, we have extensively studied the physically unclonable function concept through an elaborate overview and discussion of known constructions. This overview provides a deep exploration of the great diversity of different PUF proposals, each with its own implementation details, practical considerations, security characteristics and performance results. The listing and comparison of these features, which by itself already serves as a convenient reference work, will be of great value when trying to determine a set of common defining properties of PUFs in Chap. 3. We also propose an interesting classification of PUFs into intrinsic and non-intrinsic based on certain practical qualities of their construction. It is argued why intrinsic PUFs are considered advantageous with regard to security and cost-efficiency. In the presented overview, as is the case in PUF-related literature, the focus is therefore on intrinsic PUFs.

2.6.2 Insight into PUF Constructions

The presented overview also provides a lateral insight into PUF implementation techniques and their effect on the quality of the PUF behavior. When we focus on the intrinsic PUF proposals, a number of interesting observations can be made:

- Most intrinsic PUF constructions deploy a *differential measurement* technique, e.g. the arbiter PUF and the ring oscillator PUF, and all of the PUFs based on bistability. This turns out to be beneficial, both for the uniqueness and the reliability of their responses. Considering uniqueness, a (small) differential structure is often much more strongly affected by the locally random influence of process variations, and less prone to exhibit bias due to deterministic variations which mostly manifest at larger scales. For reliability, it is wellknown that a differential measurement exhibits a much lower sensitivity to measurement noise and is able to even completely compensate the effect of some external conditions on a measured value.
- For a given PUF construction, the quality of its PUF behavior can be enhanced in one of two ways to amplify uniqueness and reduce noise: (i) by physically tweaking the implementation details, and (ii) by algorithmically post-processing the measured values into responses.
- For physical enhancements of a given PUF construction, it is a rule of thumb that the quality of a construction's PUF behavior is directly influenced by the degree of low-level physical control one has over the implementation technology. The reasoning behind this is that the source of the PUF behavior is always of a purely physical nature, and the closer to bare physics one can design a construction, the more possibilities one has of accurately capturing this behavior. This is clear from the results on the mixed-signal-based intrinsic PUF constructions in Sect. 2.4.7: by evaluating the PUF behavior at the analog level, one is able to build more reliable and more unique PUFs. However, on the down side, the development effort also rises exponentially when designing an implementation at increasingly lower levels of physical abstraction. Additionally, many standard manufacturing flows do not provide low levels of physical manufacturing detail.
- Simple algorithmic improvements such as masking of unreliable responses and majority voting over many responses to improve reliability and/or uniqueness are in some proposals considered an inherent part of the PUF construction. As will become clear in the following chapters, algorithmic post-processing of PUF responses, to make them fit for being used in an application, can take very elaborate forms and is typically not considered as part of the PUF construction, but as a separate primitive. Algorithmic improvements nearly always imply a trade-off between the efficiency and/or performance of the PUF construction and the quality of its output, e.g. to obtain a number of PUF-derived bits with a very high reliability level, the required number of actual PUF response bits needed can be tenfold or even much more.

Chapter 3
Physically Unclonable Functions: Properties

3.1 Introduction

3.1.1 Motivation

In Chap. 2, we introduced the PUF concept and illustrated it by means of an extensive enumeration of exemplary constructions which have been proposed over the years. From this list it is clear that the term 'PUF' has been used, in printed publications but even more so in colloquial speech, as a label for a wide variety of different constructions. However, intuitively it is clear that all these constructions share a number of specific properties. When properly and unambiguously described, these properties allow us to *define* a PUF, i.e. to identify the specific attributes which make us intuitively label certain constructions as PUFs and others not.

In many publications which introduce a new PUF construction, a definition for the more general PUF concept is attempted, using varying degrees of formalism. While often fitting for the construction proposed in the same publication, most of these definitions run into problems when applied to the wider group of different PUF constructions, as discussed in Chap. 2:

- Some proposed definitions are too strict since they clearly exclude a number of constructions which are labelled as PUFs, e.g. the early definition of a physical one-way function as proposed by Pappu [104] includes a one-wayness property, but almost none of the PUF constructions proposed afterwards meet this very strict condition.
- Other definitions are too loose, i.e. they apply equivalently to constructions which are generally not considered as PUFs, e.g. true random number generators.
- Many of the proposed definitions are an ad hoc listing of the perceived qualities of the simultaneously proposed new PUF construction, and lack a degree of generalization.

These issues, combined with the fact that there are nearly as many different PUF definitions as there are PUF constructions, have caused confusion and even lead to problematic situations. One particular problem which occurs regularly is that some

R. Maes, *Physically Unclonable Functions*, DOI 10.1007/978-3-642-41395-7_3,
© Springer-Verlag Berlin Heidelberg 2013

of these ad hoc properties which often appear in PUF definitions are quickly generalized to *all* PUFs, whereas many proposed PUF constructions do in fact not meet
them. It becomes even worse when some recurring properties are in practice even
assumed for a newly proposed PUF construction, without ever actually verifying
if the construction meets them. It goes without saying that this can lead to disastrous failures when security-sensitive systems using such a PUF construction rely
on these properties.

Most of these PUF definitions are moreover of an informal nature. While not
problematic by itself, this does cause issues when PUFs are deployed in formal
systems such as cryptographic algorithms and protocols. Designers of such formal
systems are typically not familiar with the physical intricacies of a PUF construction
and need to rely on a strictly defined formal model of the PUF. On the other hand,
it turns out to be particularly difficult to capture the physical behavior in an unambiguous and meaningful formal description. Either the description is not formal
enough, which significantly reduces the usability for a formal designer, or it is too
formal, which makes it impossible for practice-oriented PUF developers to evaluate
the strict formal conditions for their PUF construction. Especially in the latter case,
there is the risk of introducing a model which is picked up in the formal world, but
which has no realistic connection to actual implementations any more.

3.1.2 Chapter Goals

Whereas Chap. 2 was intended as an exploration of the expanding field of physically
unclonable functions, the goal of this chapter is to introduce a classification in this
large and widely differing collection of PUF proposals based on the algorithmic
properties of their challenge-response behavior. In this chapter we aim to:

- Give a detailed overview of significant properties which have, at one point or
  another, been attributed to PUFs, and propose a semi-formal definition of these
  properties based on their intuitive description.
- Make a comparative analysis of the defined properties on a representative subset of PUF proposals in order to distinguish between defining and nice-to-have
  characteristics. This analysis will yield a much more tactile definition of what we
  have intuitively called a PUF in Chap. 2.
- Discuss a formal framework for deploying PUFs in formal security models, which
  is partially based on this semi-formal study. The objective is to develop a framework which is formal enough to allow meaningful security reductions, but at the
  same time sufficiently realistic and flexible to actually demonstrate the formally
  defined properties for real PUF implementations.

3.1.3 Chapter Overview

In Sect. 3.2, we do a study on the many different PUF properties which have been
proposed over time. By means of a comparative analysis over a representative set

of PUF constructions, we identify which properties are defining and which are only nice to have. This study is an extended and updated version of our earlier work in [83, Sect. 4]. In Sect. 3.3, we propose a set of formal definitions of the most important PUF properties, which is intended as a carefully balanced interface between practical PUF developers and theorists. This section is based on our work in [3]. We conclude this chapter in Sect. 3.4.

3.2 A Discussion on the Properties of PUFs

In this section, we start by listing a number of properties which are sensible to assess for PUFs, and most of them have therefore been attributed at one or more occasions to particular PUF constructions. We define these properties in a semi-formal way using the notation introduced in Sect. 2.2. The definitions are semi-formal in the sense that, while attempting to be as unambiguous as possible, we refrain from introducing a plethora of quality parameters which would make the notation needlessly complex. In that respect, we use informal qualifiers like *easy* and *hard*, *small* and *large*, and *high* and *low* to express the bounds imposed by most properties.

To avoid confusion, we want to state explicitly that at this moment we only define a number of properties, and we make no claims yet as to which PUF constructions satisfy which properties, or as to which properties are naturally implied for all PUFs. Such an analysis is only done in Sect. 3.2.9 based on a representative subset of proposed PUF constructions and is discussed in Sect. 3.2.10.

3.2.1 Constructibility and Evaluability

The notion of evaluability of a PUF construction is a purely practical and rather basic consideration expressing the fact that the required effort to obtain a meaningful outcome of a PUF instance should be feasible. Before defining evaluability, we first want to introduce the even more basic notion of constructibility, i.e. the condition that it is actually possible to produce instantiations of a particular PUF design.

Constructibility

Definition 5 A PUF class $\mathcal{P}$ is constructible if it is easy to invoke its Create procedure and produce a random PUF instance puf $\leftarrow \mathcal{P}.\text{Create}(r^C \stackrel{\$}{\leftarrow} \{0, 1\}^*)$.

It is hard to discuss the remaining properties for proposals which have no feasible instantiations. Constructibility is therefore a conditio sine qua non for evaluability, and by extension for all of the following properties listed here. The qualifier 'easy' in the definition is context-dependent. Since PUFs are physical objects, their constructibility requires at least that they be possible within the laws of physics. From a more practical viewpoint, 'easy' relates to the cost of producing an instance of a particular PUF class. An important detail in the definition of constructibility is that it is merely easy to construct a *random* PUF instance, i.e. without any specific requirements on its challenge-response behavior, whereas constructing a *specific* PUF instance can be infeasibly hard. In Sect. 3.2.4, we will discuss why it is even desirable that this is hard.

Evaluability

Definition 6 A PUF class $\mathcal{P}$ exhibits evaluability if it is constructible, and if for any random PUF instance $\mathsf{puf} \in \mathcal{P}$ and any random challenge $x \in \mathcal{X}_\mathcal{P}$ it is easy to evaluate a response $y \leftarrow \mathsf{puf}(x).\mathsf{Eval}(r^\mathsf{E} \overset{\$}{\leftarrow} \{0, 1\}^*)$.

Since the following properties all deal with the challenge-response behavior of PUF instances, it is hard to discuss the meaningfulness of constructions which are not evaluable. The 'easiness' expressed in the definition is again context-dependent. In a theoretical treatise, this typically points to some variant of 'in polynomial time and effort'. Practically however, an easy evaluation means an evaluation which is possible within the strict timing, area, power, energy and cost budget imposed by the application. From this point of view, an evaluation which is easy for one application could be infeasible for another one.

3.2.2 Reproducibility

The first property of a PUF's challenge-response behavior we will discuss is its reproducibility, which is technically also of a practical nature. However, as will become clear later on, it also has strong repercussions on the attainable security parameters of PUF-based applications.

Definition 7 A PUF class $\mathcal{P}$ exhibits reproducibility if it is evaluable, and if

$$\Pr\left(D_\mathcal{P}^{\mathsf{intra}} \text{ is small}\right) \text{ is high.}$$

Reproducibility is defined with respect to the distribution of the response intra-distance of the entire PUF class, i.e. considering evaluations of random challenges on random PUF instances. This means that with high probability, responses resulting from evaluating the same challenge on the same PUF instance should be similar, i.e. close in the considered distance metric. If the evaluation condition (α) has an impact on the responses, the definition is extended by considering the response intra-distance under condition α: $D_{\mathcal{P};\alpha}^{\mathsf{intra}}$. The qualifiers 'small' and 'high' in the definition are context-specific. Whether an intra-distance is 'small' is generally determined in relation to similar notions in other properties such as uniqueness (cf. Definition 8), e.g. as made explicit in the definition of identifiability (cf. Definition 9). How 'high' the probability needs to be typically follows from the application requirements.

3.2.3   Uniqueness and Identifiability

The most basic security-related property of PUFs is uniqueness: the observation that a PUF response is a measurement of a random and instance-specific feature.

Uniqueness

Definition 8   A PUF class $\mathcal{P}$ exhibits uniqueness if it is evaluable, and if

$$\Pr\left(D_{\mathcal{P}}^{\text{inter}} \text{ is large}\right) \text{ is high.}$$

In the same way as reproducibility, uniqueness is defined with respect to the distribution of the response inter-distance random variable of the entire PUF class, i.e. considering evaluations of random challenges on random pairs of PUF instances. This means that, with high probability, responses resulting from evaluating the same challenge on different PUF instances should be dissimilar, i.e. far apart in the considered distance metric. Uniqueness is generally assessed at nominal operating conditions; hence there is no need to extend this definition to varying evaluation conditions. The qualifiers 'large' and 'high' are again context-specific.

Identifiability

When a PUF class exhibits both reproducibility and uniqueness, it follows that its PUF instances can be identified based on their responses. We express this in the separate property of identifiability.

Definition 9   A PUF class $\mathcal{P}$ exhibits identifiability if it is reproducible and unique, and in particular if

$$\Pr\left(D_{\mathcal{P}}^{\text{intra}} < D_{\mathcal{P}}^{\text{inter}}\right) \text{ is high.}$$

Identifiability expresses the fact that responses (to the same challenge) coming from a single PUF instance are more alike than responses coming from different instances. This means that, using their response evaluations, instances of the PUF class can state a static identity which is with high probability unique. The details of how such an identification scheme can be implemented are given in Sect. 5.2. The extent to which a PUF class is identifiable is often quickly estimated based on experimental results, by comparing the average observed intra- and inter-distances and demonstrating that $\mu_{\mathcal{P}}^{\text{intra}} \ll \mu_{\mathcal{P}}^{\text{inter}}$. From this and the following definitions, it is also evident why the distributions of $D_{\mathcal{P}}^{\text{inter}}$ and $D_{\mathcal{P}}^{\text{intra}}$ play a pivotal role in the assessment of the usability of a particular PUF class, and why it is important to get accurate information about these distributions from experimental statistics.

3.2.4 Physical Unclonability

Assume an adversary which has control over the creation procedure of a PUF class, i.e. it can influence the conditions, parameters and randomness sources of $\mathcal{P}$.Create to a certain (feasible) extent. When considering identification based on PUF responses in the presence of such an adversary, a stronger argument is required to ensure that all PUF-based identities are unique with high probability. This is because this adversary can use its control over the instance creation process to attempt to produce two PUF instances which are more alike than one would expect based on the uniqueness property. To avoid this, one would like to *enforce* the uniqueness property, i.e. to ensure that the uniqueness property is met, even in the presence of such an adversary. This is what we call *physical unclonability*.

Definition 10 A PUF class $\mathcal{P}$ exhibits physical unclonability if it is evaluable, and if it is hard to apply and/or influence the creation procedure $\mathcal{P}$.Create in such a way as to produce two distinct PUF instances puf and puf$' \in \mathcal{P}$ for which it holds that

$$\Pr\left(\mathbf{dist}\left[Y \leftarrow \mathsf{puf}(X); Y' \leftarrow \mathsf{puf}'(X)\right] < D_{\mathcal{P}}^{\text{inter}}(X)\right) \text{ is high,}$$

for $X \leftarrow \mathcal{X}_{\mathcal{P}}$. In extremis, it should be very hard to produce two PUF instances for which it holds that

$$\Pr\left(\mathbf{dist}\left[Y \leftarrow \mathsf{puf}(X); Y' \leftarrow \mathsf{puf}'(X)\right] > D_{\mathcal{P}}^{\text{intra}}(X)\right) \text{ is low.}$$

The qualifier 'hard' in this definition reflects the physical and technical difficulties (or impossibility) in creating such a PUF instance pair. These difficulties need to be evaluated with respect to the technical capabilities of the adversary, which ultimately is a function of its expertise and its equipment budget. Note that the difficulty of producing a non-unique PUF instance pair, as described in the definition, implies the difficulty of producing a single PUF instance which is more alike to a given PUF instance than expressed by the uniqueness property. When combined with constructibility, physical unclonability can be summarized as: *it is easy to create a random PUF instance, but hard to create a specific one.*

A PUF class which exhibits physical unclonability has the interesting security advantage that even the genuine manufacturer of PUF instances has no way of breaking the uniqueness property. This means that one does not need to *trust* the manufacturer to make sure every PUF instance is unique with high probability, since this is implied by the physical unclonability of the PUF class. This advantage of 'physically unclonable PUFs' is called *manufacturer resistance*.

3.2.5 Unpredictability

Many applications of PUFs rely on their challenge-response functionality, i.e. the ability to apply a challenge and to receive a random response in reply. In that respect, uniqueness, and by extension physical unclonability, are often not sufficient to ensure security. One also requires unpredictability between responses on a single PUF instance, i.e. unobserved responses remain sufficiently random, even after observing responses to other challenges on the same PUF instance.

Definition 11 A PUF class $\mathcal{P}$ exhibits unpredictability if it is evaluable, and if it is hard to win the following game for a random PUF instance $\mathsf{puf} \in \mathcal{P}$:

- In a learning phase, one is allowed to evaluate puf on a limited number of challenges and observe the responses. The set of evaluated challenges is $\mathcal{X}'_{\mathcal{P}}$ and the challenges are either randomly selected (weak unpredictability) or adaptively chosen (strong unpredictability).
- In a challenging phase, one is presented with a random challenge $X \leftarrow \mathcal{X}_{\mathcal{P}} \setminus \mathcal{X}'_{\mathcal{P}}$. One is required to make a prediction Y_{pred} for the response to this challenge when evaluated on puf. One does not have access to puf, but the prediction is made by an algorithm $\mathsf{predict}$ which is trained with the knowledge obtained in the learning phase: $Y_{\mathsf{pred}} \leftarrow \mathsf{predict}(X)$.
- The game is won if

$$\Pr\big(\mathbf{dist}\big[Y_{\mathsf{pred}} \leftarrow \mathsf{predict}(X); Y \leftarrow \mathsf{puf}(X)\big] < D_{\mathcal{P}}^{\mathsf{inter}}(X)\big) \text{ is high.}$$

Note the similarity in the expressions involving the distribution of the inter-distance in this definition and the definition of physical unclonability. However, instead of considering the distance to a second created PUF instance puf', here we consider the distance to a prediction algorithm $\mathsf{predict}$ which is trained on an earlier observed set of challenges and responses on the same PUF instance. We use a game-based description for unpredictability to avoid having to put any restrictions on the prediction algorithm, i.e. unpredictability is defined with respect to the best conceivable prediction algorithm which can be built, trained and evaluated within the capabilities of the adversary. In the best case, one can show that responses to different challenges are completely independent, which means they cannot be predicted by any prediction algorithm. However, such a strong quality can rarely be proven for a PUF construction, and at best one can assume it for certain PUFs based on a physical motivation.

For other PUF constructions, responses are not independent and their unpredictability relies on the computational difficulty of constructing, training and evaluating an appropriate prediction algorithm. In that case, the extent of unpredictability can only be estimated in relation to the currently best-known modeling attack, since there is no guarantee that no better attacks exist. This is similar to the situation for most cryptographic symmetric primitives, e.g. a block cipher like AES, where a primitive is only as secure as indicated by the currently best-known attack. Kerckhoffs' principle typically also applies to PUFs, i.e. an adversary has full knowledge about the design and implementation details of a particular PUF construction, except for the instance-specific random features introduced during the creation process. The efficiency of a practical model building attack aimed at breaking the unpredictability of a PUF is typically expressed by their prediction accuracy as a function of the size of the training set, e.g. as done in the description of modeling attacks on arbiter PUFs in Sect. 2.4.1.

3.2.6 Mathematical and True Unclonability

In the definition of unpredictability, an adversary is restricted to learning a limited number of (possibly random) challenge-response pairs which it uses to train its prediction algorithm. This is typically the case in a challenge-response-based protocol where the adversary eavesdrops on the protocol communications. However, a stronger adversarial model needs to be considered when the adversary has unlimited physical access to a PUF instance. This means it can learn as many challenge-response pairs as it is capable of storing and possibly even make useful observations of the PUF instance beyond the challenge-response functionality.

Definition 12   A PUF class $\mathcal{P}$ exhibits mathematical unclonability if it is unpredictable, even if there is no limit on the access to the PUF instance during the learning phase (as described in Definition 11), besides one's own capacities.

Mathematical unclonability is hence the extension of unpredictability to an adversary with unlimited physical access to a PUF instance. It is therefore evident that mathematical unclonability implies unpredictability.

A direct condition for a PUF class $\mathcal{P}$ to be mathematically unclonable is that its challenge set $\mathcal{X}_\mathcal{P}$ be very large, preferably exponential in some construction parameter of $\mathcal{P}$. If this is not the case, an adversary with unlimited physical access to a PUF instance can evaluate the complete challenge set and store the observed responses in a table. This table then serves as a perfect prediction model of the considered PUF instance and the unpredictability property is broken (technically, there are no challenges left to play the challenging phase of the unpredictability game with). The same argument holds if the challenge set is large, but if a near-perfect response prediction algorithm can be trained based on a small subset of these challenges.

True Unclonability

We have defined two different notions of unclonability: physical and mathematical unclonability. Both describe a property with the same objective, i.e. it is hard to *clone* a PUF instance, but from completely different perspectives. Physical unclonability deals with actual physical clones of PUF instances, whereas mathematical unclonability deals only with cloning the challenge-response behavior of a PUF instance. For a PUF class to exhibit true unclonability, both properties need to be met.

Definition 13 A PUF class $\mathcal{P}$ exhibits true unclonability if it is both physically and mathematically unclonable.

3.2.7 One-Wayness

We describe a one-wayness property for PUFs similar to the one in the definition of physical one-way functions as proposed by Pappu [104].

Definition 14 A PUF class $\mathcal{P}$ exhibits one-wayness if it is evaluable, and if given a random PUF instance $\mathsf{puf} \in \mathcal{P}$, there exists no efficient algorithm $\mathsf{invert}_{\mathsf{puf}}$: $\mathcal{Y}_{\mathcal{P}} \rightarrow \mathcal{X}_{\mathcal{P}}$ which is allowed to evaluate puf a feasible number of times and for which it holds that

$$\Pr\big(\mathbf{dist}\big[Y \leftarrow \mathsf{puf}(X); Y' \leftarrow \mathsf{puf}\big(\mathsf{invert}_{\mathsf{puf}}(Y)\big)\big] > D_{\mathcal{P}}^{\mathsf{intra}}(X)\big) \text{ is low,}$$

for $X \leftarrow \mathcal{X}_{\mathcal{P}}$.

Hence, given a PUF instance and a random response of that instance, there exists no efficient inversion algorithm acting on the PUF instance, that finds a challenge that would produce a response close to the given response. This definition resembles the classic definition of a one-way function in theoretical cryptography, but takes the unreliability and the uniqueness of a PUF instance into account. However, the notion of one-wayness is somewhat ambiguous for PUF constructions since, besides depending on the actual algorithmic complexity of inverting a PUF instance, it also depends very much on the attainable sizes of the challenge and response sets of the PUF constructions. For PUF constructions with a small challenge set, one-wayness is not achievable since the inversion algorithm can easily evaluate every possible challenge and perform an inverse table lookup to invert a given response. On the other hand, if the response set is small, the inversion algorithm can evaluate random challenges on the PUF instance and will quickly encounter one which inverts a given response.

3.2.8 Tamper Evidence

Tampering is the act of permanently altering the physical integrity of a system, e.g. of a PUF instance, with the intent of modifying its operation in an unauthorized and possibly harmful manner. It is hence a type of physical transformation which we denote as $\mathsf{puf} \Rightarrow \mathsf{puf}'$ to make clear that physical changes were made. Directed tampering represents a powerful attack against security implementations. It can be used to remove or bypass protection mechanisms and leave the implementation vulnerable, or to obtain information about sensitive internal values and parameters. Protection against tampering is a matter of detecting tampering and providing an appropriate reaction, e.g. clearing confidential data and/or blocking all functionality. In order to detect tampering, a security system needs to have some level of tamper evidence, i.e. a tampering attempt will have an unavoidable and measurable impact on the system. Certain types of PUF constructions, which rely on sensitive measurements of random physical features of an instance, cannot be physically tampered with without significantly changing their challenge-response behavior.

Definition 15 A PUF class $\mathcal{P}$ exhibits tamper evidence if it is evaluable, and if given a random PUF instance $\mathsf{puf} \in \mathcal{P}$, any physical transformation of $\mathsf{puf} \Rightarrow \mathsf{puf}'$ has the effect that

$$\Pr\big(\mathbf{dist}\big[Y \leftarrow \mathsf{puf}(X); Y' \leftarrow \mathsf{puf}'(X)\big] > D_{\mathcal{P}}^{\mathsf{intra}}(X)\big) \text{ is high},$$

and ideally that

$$\Pr\big(\mathbf{dist}\big[Y \leftarrow \mathsf{puf}(X); Y' \leftarrow \mathsf{puf}'(X)\big] < D_{\mathcal{P}}^{\mathsf{inter}}(X)\big) \text{ is low}.$$

Informally, tamper evidence means that it is very hard to physically alter a PUF instance without having a noticeable effect on its challenge-response behavior, i.e. an effect which with high probability is larger than the unreliability of the PUF as expressed by the distribution of $D_{\mathcal{P}}^{\mathsf{intra}}$. Ideally, such an alteration even causes the PUF instance to become a completely different one, i.e. the effect on its challenge-response behavior is indiscernible from replacing the PUF instance with a different unique instance as expressed by the distribution of $D_{\mathcal{P}}^{\mathsf{inter}}$.

In a sense, tamper evidence is an orthogonal property to physical unclonability. Physical unclonability states that it is very hard to make two distinct PUF instances more alike than is to be expected from physically different instances. Tamper evidence on the other hand states that it is very hard to physically alter a single PUF instance resulting in a different PUF instance which is more alike to the original than is to be expected from physically different instances. Tamper-evident PUFs are self-protecting in the sense that a tampering attack on their implementation unavoidably and substantially alters their responses, which generally results in a blocked functionality or a loss of secret information. Moreover, they can also be deployed as to make a larger security system tamper-resistant by encapsulating the system in a tamper-evident PUF instance.

3.2.9 PUF Properties Analysis and Discussion

We will now evaluate the properties defined in this section on a representative set of proposed PUF constructions described in Chap. 2, as well as on a number of non-PUF reference cases. This analysis will clarify which of these properties are met by all PUF proposals, which are only met by a few or by none at all, and most importantly, which subset of properties differentiates the PUF and non-PUF constructions.

Representative Subset of Constructions

We have selected a representative set of both non-intrinsic and intrinsic PUF constructions to do this comparative analysis on. We have only selected proposals for which actual implementations have been done and for which experimental data is available. The PUF constructions we will consider are:

1. The optical PUF as proposed by Pappu et al. [105].
2. The coating PUF as proposed by Tuyls et al. [147].
3. The simple arbiter PUF as proposed by Lee et al. [75].
4. The feed-forward arbiter PUF as proposed by Lee et al. [75].
5. The XOR arbiter PUF as proposed by Majzoobi et al. [94].
6. The basic ring oscillator PUF as proposed by Suh and Devadas [136].
7. The enhanced ring oscillator PUF as proposed by Maiti et al. [92].
8. The SRAM PUF as proposed by Guajardo et al. [45].
9. The latch, flip-flop, butterfly and buskeeper PUFs as respectively proposed by Su et al. [135], Maes et al. [84], Kumar et al. [72] and Simons et al. [132]. It is clear that these PUFs will have identical properties since they only differ slightly in their construction details. Therefore we treat them simultaneously.
10. The glitch PUF as proposed by Shimizu et al. [129].
11. The bistable ring PUF as proposed by Chen et al. [22].

To be able to differentiate between PUFs and non-PUF constructions which are used to achieve similar objectives, we have selected a representative set of non-PUF reference constructions. In order to be able to assess the properties of this section on these non-PUF references, we explicitly state what we consider to be the challenge and the response. The non-PUF reference cases we consider are:

1. A true random number generator or TRNG. A TRNG is a process which generates a stream of uniformly random numbers based on measurements of a dynamically random physical process. A TRNG's output is *truly random* since it results from a non-deterministic physical process, as opposed to the output of a mathematical algorithm, known as a seeded *pseudo-random* number generator, which only appears random but can in fact be deterministically calculated if the seed is known. For easy comparison, we say the single challenge of the TRNG is a request for a random number and the response is a fixed-length random output string.
2. A very simple, unsecured radio-frequency identification (RFID) scheme. In this most basic form, an RFID tag is nothing more than a small non-volatile memory which upon deployment is programmed with a unique identifier string. When triggered by a reader, the tag broadcasts this string to identify itself. For easy comparison, we say the single challenge of this RFID scheme is the reader's trigger and the response is the identifier string broadcasted by the tag.
3. An implementation of a secure (unkeyed) cryptographic hash function. We say the challenge is a (fixed-length) message string and the response is the resulting hash digest of that message.

4. An implementation of a secure cryptographic block cipher, with a randomly generated encryption key which is programmed in a non-volatile memory in the implementation. We say the challenge of this block cipher is a single block of message data and the response is the resulting block of ciphertext data.
5. An implementation of a secure cryptographic public-key signature algorithm, with a randomly generated public/private key pair of which the private key is programmed in a non-volatile memory in the implementation. We say the challenge of this signature algorithm is a (fixed-length) message string and the response is the resulting signature on this message.

Comparative Analysis

For every construction listed above, we have assessed whether it meets the different properties discussed in this section. The resulting analysis is shown in Table 3.1. We make four distinctions as to what extent a construction exhibits a certain property:

- **V**: the construction exhibits the property fully or to a large extent
- **X**: the construction does not exhibit the property
- **!**: the construction only exhibits the property under certain conditions
- **?**: it is not clear or unknown whether the construction exhibits the property

Note that the assessments presented in Table 3.1 are only a reflection of the current state of the art, to the best of our knowledge. However, due to the natural progress in mathematical and physical attacks, and in manufacturing techniques, some of these classifications can change over time. We will shortly discuss the presented results for each property and the reasoning behind certain classifications.

Constructibility   All considered constructions are constructible since for all of them known implementations exist. However, some of them require more construction effort than others. The optical and coating PUFs rely on explicitly introduced randomness during manufacturing and are therefore non-intrinsic; the other considered PUF constructions are intrinsic. The RFID scheme, the block cipher and the signature algorithm also require explicit programming of a random string.

Evaluability   All considered constructions are also evaluable since experimental results are available for all of them. However, some of them require more evaluation effort than others. The optical PUF evaluation procedure is quite elaborate, requiring a laser and a very accurate mechanical positioning system. The SRAM PUF and the flip-flop and buskeeper PUFs require a power cycle to evaluate them, since they rely on the power-up behavior of their construction.

Reproducibility   This is the first differentiating property between PUFs and non-PUFs, since it clearly distinguishes the PUF constructions from the TRNG. A TRNG is by definition not reproducible, since this would make it deterministic. The other non-PUF reference cases are perfectly reproducible, i.e. with zero intra-distance. The reproducibility of the PUF proposals varies from more than 25 % average intra-distance for the optical PUF to less than 1 % for the ring-oscillator PUF.

Table 3.1 Properties of a representative subset of PUF and non-PUF constructions

Property	Optical PUF	Coating PUF	Simple Arbiter PUF	FF Arbiter PUF	XOR Arbiter PUF	Basic Ring Oscillator PUF	Enhanced Ring Oscillator PUF	SRAM PUF	Flip-flop / Butterfly / Latch / Buskeeper PUF	Glitch PUF	Bistable Ring PUF	TRNG Output	RFID Broadcast	Cryptographic Hash Function	Block Cipher Encryption	Public-key Signature
Constructibility	✓	✓	✓	✓	✓	✓	✓	✓	✓	✓	✓	✓	✓	✓	✓	✓
Evaluability	✓	✓	✓	✓	✓	✓	✓	✓	✓	✓	✓	✓	✓	✓	✓	✓
Reproducibility	✓	✓	✓	✓	✓	✓	✓	✓	✓	✓	✓	✗	✓	✓	✓	✓
Uniqueness	✓	✓	✓	✓	✓	✓	✓	✓	✓	✓	✓	✓	✓	✗	✓	✓
Identifiability	✓	✓	✓	✓	✓	✓	✓	✓	✓	✓	✓	✗	✓	✗	✓	✓
Physical Unclonability	✓	✓	✓	✓	✓	✓	✓	✓	✓	✓	✓	✓	✗	✗	✗	✗
Unpredictability	✓	✓	–	–	?	✗	?	✓	✗	?	?	✓	✗	✗	✓	✓
Mathematical Unclonability	✓	✗	✗	✗	?	✗	?	✗	✗	?	?	✓	✗	✗	–	–
True Unclonability	✓	✗	✗	✗	?	✗	?	✗	✗	?	?	✓	✗	✗	✗	✗
One-wayness	✓	✗	✗	✗	✗	✗	✗	✗	✗	✗	✗	✓	✗	✓	✓	✗
Tamper Evidence	✓	✓	?	?	?	?	?	?	?	?	?	?	–	✗	–	–

Legend:

✓   construction exhibits property

✗   construction does not exhibit property

?   it is unknown/untested whether construction exhibits property (more research required)

–   construction exhibits property only under certain conditions

Uniqueness All PUF proposals exhibit uniqueness with average inter-distances of merely 23 % and 38 % respectively for the simple and the feed-forward arbiter PUFs, and very close to 50 % for the other constructions. The TRNG evidently also exhibits ideal uniqueness. Uniqueness is a differentiating feature between PUFs and hash functions. Because the considered hash function has no random elements, every implementation instance will exhibit exactly the same challenge-response behavior. The other three non-PUF constructions are unique if and only if they have been programmed with a unique bit string, which is true with high probability in the regular manufacturing process.

Identifiability This is the combination of reproducibility and uniqueness. TRNGs and hash functions do not meet this property, respectively for not being reproducible and not being unique. All other constructions exhibit identifiability, with for most of them even a large separation between the distributions of intra- and inter-distance, allowing an unambiguous identification based on their responses.

Physical Unclonability As expected, physical unclonability is the great divider between PUF and non-PUF constructions. The uniqueness of all PUF proposals results from random physical processes during their manufacturing process, either implicitly or explicitly, which are so complex that it is considered infeasible to have any meaningful influence on them. Moreover, for all considered PUF constructions, these random processes take effect at microscopic and (deep) submicron levels, which makes it even more technically infeasible to analyse them or exert any control over them. This is in great contrast to the uniqueness of the RFID scheme, the block cipher and the signature algorithm which is based on the programming of a relatively short random bit string. For an adversary which controls the manufacturing process, and hence this programming step, it is not at all difficult to program more than one instance with the same string. For this reason, these constructions are not considered physically unclonable.

Unpredictability Following the modeling attacks on simple and feed-forward arbiter PUFs as discussed in Sect. 2.4.1, these two constructions only remain unpredictable as long as an adversary does not learn enough challenge-response pairs to accurately train its model. The actual number of unpredictable responses depends on the details of the implementation and on the state of the art in modeling attacks, but this can be very limited: modeling attacks have been presented which achieve better than random accuracy after less than 100 training responses and improve to near-perfect accuracy when more responses are learned. For the XOR arbiter PUF with a sufficient number of XOR-ed arbiters (≥ 5), no effective modeling attacks are yet known, and the same holds for the enhanced ring oscillator PUF, the glitch PUF and the bistable ring PUF, although for none of these four PUFs can strong claims of independence between response bits be made. For the optical PUF, reasonable arguments are presented by Škorić et al. [133], Tuyls et al. [146] that modeling attacks are computationally infeasible. For the remaining PUFs, response bits are produced by physically distinct elements, which is a strong motivation to assume that different responses are independent and hence inherently unpredictable. Consequently, no modeling attacks are known for these PUFs. All the non-PUF constructions except for the RFID scheme and the hash function exhibit unpredictability, the TRNG because of random physical influences and the block cipher and the signature scheme because of a secret element and computational complexity arguments. The RFID scheme is trivially predictable since it only has one single fixed response. The hash function construction contains no unique element (key) and is easily predicted.

Mathematical Unclonability   Simple and feed-forward arbiter PUF constructions are not mathematically unclonable since they are only conditionally unpredictable. The coating PUF, the ring oscillator PUF, the SRAM PUF and the four other bistable memory element PUFs do not exhibit mathematical unclonability because the size of their challenge sets is small, which means they can be fully evaluated to produce a lookup table as a trivial mathematical clone. For the XOR arbiter PUF, the enhanced ring oscillator PUF, the glitch PUF and the bistable ring PUF, no practical mathematical cloning attacks are known, but more research is required before any strong claims can be made. Škorić et al. [133], Tuyls et al. [146] argue that the optical PUF is to a large extent mathematically unclonable, since even if (hypothetically) a mathematical model of an optical PUF instance could be created, it would be computationally too complex to be evaluated. However, this turns out to be very implementation-dependent, as Rührmair [113] states that a variant of an optical PUF with reduced randomness can in fact be modeled. Finally, the block cipher and the signature algorithm are assumed to be mathematically unclonable under the condition that an adversary is not able to extract the secret key during its unlimited physical access to an instance. This implies that, besides being algorithmically secure, the implementations of these cryptographic primitives also need to be physically protected, e.g. against side-channel and fault attacks, and against reverse-engineering.

True Unclonability   This is the combination of physical and mathematical unclonability. The optical PUF is the only one of the considered PUF constructions for which strong claims of true unclonability can be made.

One-Wayness   Besides the optical PUF, none of the other studied PUF proposals can be considered one-way since they either have a small challenge or a small response set. Pappu [104] presents strong arguments why his optical PUF construction does exhibit one-wayness, hence being labelled a physical one-way function. If the block cipher is algorithmically secure and its key remains secret, it is considered to be hard to invert. The same holds for the unkeyed hash function. A public-key signature algorithm is not guaranteed to be one-way, since everyone with knowledge of the public key can verify the signature, which possibly involves recovering the signed message.

Tamper Evidence   A certain level of tamper evidence was only experimentally demonstrated for optical PUFs by Pappu [104] and for coating PUFs by Tuyls et al. [147]. For all other proposed PUF constructions no results on tamper evidence, neither in the positive nor in the negative sense, are known. Hence no sensible claims

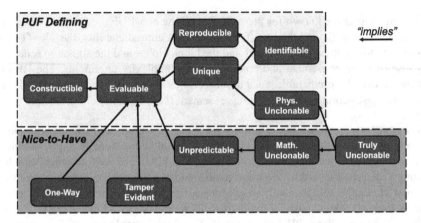

Fig. 3.1 Relations between the different described PUF properties and an indication of the PUF defining properties

on tamper evidence can be made for them. For TRNGs, tamper evidence is not well defined. The remaining non-PUF proposals are not inherently tamper-evident, but can be implemented in a tamper-evident fashion by applying tamper detection measures.

3.2.10 Discussion on PUF Properties

PUF Defining Properties

Looking at the results of the comparative property study in Table 3.1, and assuming these are representative for all PUF and non-PUF constructions, we can determine the properties which are *defining* for PUFs, i.e. which properties do all PUFs meet that distinguish them from all conceivable non-PUF constructions? The answer to this question turns out to be *identifiability*, and to a larger extent *physical unclonability*, which is in fact the enforcement of uniqueness in the presence of an adversary with control over the instance creation process. Note that these two properties imply that PUFs are also constructible, evaluable, reproducible and unique. The relations between all discussed properties, as well as an indication of the PUF defining properties, are shown in Fig. 3.1. Based on this analysis, we tentatively propose a definition of a PUF class.

Definition 16 A class of physical entities with a challenge-response functionality is called a PUF class if it exhibits identifiability (cf. Definition 9) and physical unclonability (cf. Definition 10).

In the light of this definition, we argue that the acronym 'PUF' as standing for *physically unclonable* function, i.e. with the qualifier 'physically' reflecting on 'unclonable', not on 'function', is particularly fitting for the concept. This strong emphasis on physical unclonability being the core PUF property also implicates that if at one point the physical unclonability of a particular PUF construction is disputed, e.g. due to the natural advancement of manufacturing techniques, it will cease to be a PUF.

Nice-to-Have PUF Properties

Since only identifiability and physical unclonability actually define PUFs, the remaining properties of unpredictability, mathematical and true unclonability, one-wayness and tamper evidence are merely desirable extras, but are not guaranteed for any PUF construction. In fact, there currently seems to be only one PUF proposal which meets all these properties, and that is the optical PUF as proposed by Pappu [104]. This observation, combined with the fact that it was one of the very first PUF proposals, grants the optical PUF the status of *prototype PUF*. All following PUF constructions aim to achieve as many of the desirable properties of the optical PUF as possible, but at the same time aim to provide more integrated implementations.

Since these remaining properties are only nice-to-have qualities, they cannot simply be assumed to be present in any PUF. This means that PUF designers need to present strong arguments, preferably of an experimental nature, if they claim any of these extra properties for their PUF proposal. (In fact, they also need to show identifiability and physical unclonability to demonstrate that their proposal is a PUF to begin with.) For developers of PUF-based applications, this entails that they need to state explicitly if, and to what extent, they rely on any of these nice-to-have properties, since it means that not all PUF constructions can be used for their application.

Improving PUF Properties

From Table 3.1 it is clear that mathematical unclonability, and in consequence true unclonability, and one-wayness are hard to come by for most PUF proposals. When these properties are required in a PUF-based application, they can be provided by extending the raw physical PUF instance with an algorithmic primitive exhibiting these properties, and enforcing that the resulting PUF-system only be evaluated with this primitive. This is what was defined as a *controlled PUF* in Sect. 2.5.2.

3.3 Formalizing PUFs

In Sect. 3.2, we described meaningful properties of PUFs, and after a broad comparison identified the ones which define a PUF. Based on this analysis, we proceed towards a strictly formal description of PUFs and their key properties. First, we study earlier proposed attempts at formally describing PUFs and point out how they fall short. Next we discuss our approach in setting up the framework and in Sect. 3.3.3 we introduce the basic primitives of the framework itself. Using the introduced framework, we propose formal definitions respectively for the notions of robustness, physical unclonability and unpredictability. The rationale behind the definitions of all concepts and properties in this section is to provide a meaningful formal model for both hardware engineers (developing PUFs) and cryptographers (deploying PUFs).

Background   The development of the formal framework for physical functions and the formal definitions of their properties as initially proposed in [3] and discussed in this section are the shared results of numerous fruitful discussions and an intense research collaboration between the author and Prof. Frederik Armknecht (Universität Mannheim), Prof. François-Xavier Standaert (Université catholique de Louvain), Christian Wachsmann and Prof. Ahmad-Reza Sadeghi (both Technische Universität Darmstadt).

3.3.1 Earlier Formalization Attempts

Throughout PUF literature, many authors have attempted to generalize the concept of a PUF in a more or less formal definition, mainly as a means to highlight the advantageous properties of a simultaneously proposed new PUF construction. We briefly introduce these definitions and point out why we believe none of them captures the full spectrum of proposed PUFs and their properties, either by being too restrictive, i.e. excluding certain PUFs, or by being too ad hoc, i.e. listing perceived and even assumed properties of certain PUFs instead of providing a generalizing model. A similar overview and discussion has been presented by Rührmair et al. [115]. However, we do not completely follow all their arguments and moreover point out why the new models they propose are still insufficient.

Another approach toward defining the functionality of a PUF comes from the theoretical corner. Theorists, in an attempt to deploy PUFs in their algorithms and protocols, provide rather rigid formal descriptions of PUFs on which they can built security reductions. We also discuss these proposals and their drawbacks.

Physical One-Way Functions

To the best of our knowledge, the first generalizing definition of the PUF concept is given by Pappu [104], based on the properties of his optical PUF construction. He focuses on the one-wayness property of this construction and labels it a *physical one-way function* (POWF). The first part of the definition of a POWF states that it is a deterministic physical interaction that is evaluable in constant time but cannot be inverted by a probabilistic polynomial time adversary with a non-negligible probability. The second part of the definition focusses on the unclonability of the POWF: both simulating a response and physically cloning the POWF should be hard. The POWF definition is solely based on the optical PUF, which at that time was the only known PUF. As other PUFs were introduced shortly after, it became clear that this definition was too stringent, in particular regarding the one-wayness assumption. While the optical PUF has very large challenge and response sets, many of the later introduced PUFs do not. For these constructions, one-wayness does not hold any longer since inverting such a PUF with non-negligible advantage becomes trivial, as discussed in Sect. 3.2.7. It is also noteworthy that, as pointed out by Rührmair et al. [115], for most PUF-based security applications, one-wayness is not a required condition. A final issue with the POWF definition is that it lacks any notion of noise; in fact it even describes a POWF as a deterministic interaction. This is contradicted by the fact that virtually all PUF proposals, including the optical PUF, produce noisy responses due to uncontrollable physical influences affecting a response evaluation.

Physical Random Functions

With the introduction of delay-based intrinsic PUFs, Gassend et al. [42], propose the definition of *physical random functions* to describe PUFs. In brief, a physical random function is defined as a function embodied by a physical device which is *easy to evaluate* but *hard to predict* from a polynomial number of challenge-response observations. Note that this definition replaces the very stringent one-wayness condition from POWFs with a more relaxed unpredictability condition. However, the presentation of modeling attacks on simple arbiter PUFs by Lee et al. [75] and on more elaborate arbiter PUFs by Rührmair et al. [119] demonstrate a significantly reduced unpredictability of these types of PUFs. Moreover, the later introduced memory-based intrinsic PUFs only possess at most a polynomial number of challenges and hence do not classify as physical random functions since they can be easily modeled through exhaustive readout. Finally, the definition of physical random functions as proposed by Gassend et al. [42] also does not capture the possibility of noisy responses.

Weak and Strong PUFs

With the introduction of memory-based intrinsic PUFs, Guajardo et al. [45] further refine the formal specification of a PUF. They describe PUFs as inherently unclonable physical systems with a challenge-response behavior. It is *assumed* that: (i) responses to different challenges are independent of each other, (ii) it is difficult to come up with responses which have not been observed before, and (iii) tampering with a PUF instance substantially changes its challenge-response behavior. For the first time, it is made explicit that PUF responses are observed as noisy measurements. This definition also comes with a division of *strong* and *weak* PUFs, depending on how many challenge-response pairs an adversary is allowed to obtain in order to model the PUF. If the number is exponentially large in some security parameter, the PUF is called a strong PUF; otherwise the PUF is called weak. It can be argued that some of the assumptions made in this description do not have a solid experimental basis, in particular regarding tamper evidence, which has not been tested in practice for any of the intrinsic PUF proposals. Also, strong PUFs are difficult to characterize in general, as the idea of a security parameter is specific to each PUF instance, and no practical procedure is proposed to exhibit the required exponential behavior in practice.

Rührmair et al. [115] build upon the distinction between strong and weak PUFs from Guajardo et al. [45] and redefine both notions in terms of a security game with an adversary. Weak PUFs are called *obfuscating PUFs* and are basically considered as physically obfuscated keys, as described in Sect. 2.5.1. The main statement in the definition of obfuscating PUFs is that an adversary cannot learn the key after having had access to the PUF for a limited amount of time. Strong PUFs are defined similarly, but here the adversary needs to come up with the response to a randomly chosen challenge after having had access to the PUF and a PUF oracle for some limited time. Some issues are again left unresolved in this formalization: first, despite building upon the work of Guajardo et al. [45], responses are not considered to be noisy. Next, the use of a PUF oracle in the definition of a strong PUF seems questionable. It is argued that this oracle is introduced to circumvent practical access restrictions to the PUF. However, if a PUF-based system is secured against any attacks possible with the current state of technology, the access to such an oracle is an unrealistic advantage to the adversary, which weakens the proposed definition.

PUFs and PUF-PRFs

In [2] we have introduced a first formal PUF model which starts from a theoretical application perspective, rather than from a practical construction-based perspective as most earlier proposals. In [2], we aim to use a PUF as part of a cryptographic algorithm, i.c. a block cipher, and for that goal the previously discussed definitions prove to be insufficient. We explicitly make a distinction between algorithmic and physical properties of a PUF. From the algorithmic side, a PUF is said to be a noisy function for which the distribution of responses is indistinguishable from a random distribution with a certain amount of min-entropy. From the physical side, a PUF is assumed to be physically unclonable and tamper-evident. A PUF-PRF is then defined as a PUF-based system with pseudo-random function-like qualities. We already pointed out the lack of experimental proof for tamper evidence of intrinsic PUFs in practice, and the same argument applies to this definition. The description of uniqueness and unpredictability by means of min-entropy provides convenient qualities for the theoretical application of PUFs, but as it turns out it is hard to give strong proof of min-entropy levels for actual PUF constructions. Contrarily to most of the previous definitions, PUFs are explicitly defined as noisy functions with a strictly bounded noise magnitude.

PUFs in the Universal Composition Framework

Brzuska et al. [15] continue the theoretical application viewpoint approach toward defining PUFs, and propose a very formal definition which allows them to model PUF functionality in the universal composition framework as proposed by Canetti [20]. As in our definition from [2], they define PUFs based on response distributions having a certain level of min-entropy and also incorporate a noise threshold. However, the presented formulation is utterly theoretical, which makes it unattractive for practice-oriented designers of PUF constructions. Consequentially, there is a significant probability that an actual PUF construction which lives up to this stringent definition will never be proposed. Moreover, the strong min-entropy assumptions on the PUF responses rule out many known PUF constructions and put strong restrictions on others, leading to practical inefficiency.

3.3.2 Setup of the Formal Framework

Objective

The basic objective of the formal framework presented in [3], and which is discussed in detail in the following, is to provide a workable model for both practice-oriented PUF designers as well as theory-oriented cryptographers deploying PUFs in algorithms and protocols. Ideally, the model is sufficiently realistic, capturing measurable properties of actual PUF proposals, while at the same time providing sufficient formalism and rigor to allow theoretical security reductions of systems deploying a PUF as a primitive. Such a framework could serve as an *interface* between hardware engineers and theoretical cryptographers, which would be very beneficial for the continued successful deployment of PUFs in security systems.

Approach

The approach we take is to start from a minimalistic axiomatic framework to describe *physical functions* and increment it in a flexible manner; first, by hierarchically expanding the notion of a physical function to more extensive constructions, and secondly by defining modular properties of these constructions within the framework.

The particular difficulty experienced by formal modeling attempts of PUFs is dealing with the physical aspect. It is hard to argue about the security properties of a physical object because they typically cannot be captured by classical cryptographic notions such as security parameters or computationally hard problems. To isolate this difficulty, we capture the physical aspect at the lowest formal level when we axiomatically describe the notion of a physical function. All higher-level properties can then be defined using the formalism introduced to describe physical functions. The particular properties which we consider are robustness, physical unclonability and unpredictability. Note that what we call a physical function is a very general concept, which is broader than PUFs, and physical unclonability is merely one possible property of a physical function.

3.3.3 Definition and Expansion of a Physical Function

Physical Function (pf)

A physical function pf consists of a physical component p and an evaluation procedure $\text{Eval}_{a_{ev}}$. A physical component can be physically stimulated, resulting in a measurable effect. The evaluation procedure $\text{Eval}_{a_{ev}}$ translates the physical stimulus and resulting measurement into digital forms, respectively called the challenge x and the response y of the physical function. The exact challenge-response behavior of a pf is determined by both the static and the dynamical physical states of its physical component, and by an evaluation parameter a_{ev} which controls the challenge and response translation, e.g. the quantization step size of an analog measurement.

Definition 17 A physical function pf is a probabilistic procedure

$$\text{pf}_{p,a_{ev}} : \mathcal{X} \to \mathcal{Y},$$

consisting of an evaluation procedure Eval acting upon a physical component p:

$$y \leftarrow \text{pf}_{p,a_{ev}}(x) = \text{Eval}_{a_{ev}}(p; x).$$

When the physical component and the evaluation parameter are clear from the context, we simply write pf instead of $\text{pf}_{p,a_{ev}}$.

Extraction Algorithm (Extract)

A physical function is not a function in the classical sense since, when challenged with the same challenge x twice, it may produce different responses. This is an effect of the response representing a measurement of a physical component whose physical state is partially dynamic, e.g. as a result of non-deterministic random physical noise during the measurement. However, for many applications this is an undesirable feature of a physical function, and it is dealt with by an appropriate extraction algorithm $\text{Extract}_{a_{ex}}$ for which it is possible to guarantee the reproducibility of its output. Many instantiations of extraction algorithms exist, including the seminal *fuzzy extractor* as proposed by Dodis et al. [32, 33]. As with physical functions, we describe extraction algorithms as generically as possible to allow the greatest possible flexibility of the framework. An extraction procedure extracts an output $z \in \mathcal{Z}$ from a response y of a pf, and in the process it can also consume and/or produce some additional side information which we call *helper data* and denote as $w \in \mathcal{W}$. We also introduce an extraction parameter a_{ex} which is used to exactly specify all the deployment details of the extractor.

Definition 18 An extraction algorithm Extract is a probabilistic procedure

$$\text{Extract}_{a_{ex}} : \mathcal{Y} \times \mathcal{W} \to \mathcal{Z} \times \mathcal{W},$$

which operates in one of two modes depending on the format of the presented helper data:

$$[\text{setup}] \qquad (z, w) \leftarrow \text{Extract}_{a_{\text{ex}}}(y, \epsilon),$$

$$[\text{reconstruction}] \quad (z', w' = w) \leftarrow \text{Extract}_{a_{\text{ex}}}(y', w \neq \epsilon),$$

with ϵ denoting the empty string.

When the extraction parameters are clear from the context, we simply write Extract instead of $\text{Extract}_{a_{\text{ex}}}$.

In setup mode, when no helper data is presented as an input ($w = \epsilon$), the extraction algorithm produces an output z and helper data w. In reconstruction mode, another possibly noisy evaluation of the response y' is presented together with helper data w which was produced in an earlier setup mode of the extractor. The extraction algorithm re-extracts the output z' and additionally outputs the unchanged helper data w. The power of most extraction algorithms is that, under certain conditions on the pf response distribution, they succeed in recreating exactly the same output in both setup and reconstruction mode: $z = z'$, given that the helper data generated by the setup mode is used during reconstruction mode. Besides this reconstruction property, an extraction algorithm can also provide guarantees about the randomness of its output z. The actual implementation of the extraction algorithm, as is the case for the physical function, is left up to the practical developer. The generic nature of the definition allows a wide variety of extractor implementations, including using no extractor at all by making it the identity function.

Physical Function System (pfs)

In many application scenarios, the use of an extraction algorithm is indispensable, and by consequence a user will only be aware of the challenge provided to the physical function and the output generated by the extractor. The existence of an intermediate physical function response is transparent to him. Additionally, the relevant security notions in such a scenario will be determined by the combination of both the used physical function and the deployed extraction algorithm. For these reasons, it makes sense to abstract away the separate notions of a physical function and an extraction algorithm and consider their combination as a single building block. We call such a combination a *physical function system* pfs.

Definition 19 A physical function system pfs is a probabilistic procedure

$$\text{pfs}_{p,a_{\text{ev}},a_{\text{ex}}} : \mathcal{X} \times \mathcal{W} \to \mathcal{Z} \times \mathcal{W},$$

consisting of the concatenation of a physical function (cf. Definition 17) with an extraction algorithm (cf. Definition 18):

$$(z, w') \leftarrow \text{pfs}_{p,a_{\text{ev}},a_{\text{ex}}}(x, w) = \text{Extract}_{a_{\text{ex}}}\big(\text{pf}_{p,a_{\text{ev}}}(x), w\big).$$

When the physical component and the parameters are clear from the context, we write $\mathsf{pfs}(x, w)$ instead of $\mathsf{pfs}_{\mathsf{p},a_{\mathsf{ev}},a_{\mathsf{ex}}}(x, w)$.

We believe that, from a theoretical perspective, it is easier to reason about physical function systems than to deal with the technical peculiarities of physical functions and extractors. Therefore we will define formal properties over physical function systems, rather than over physical functions, and we only refer to the underlying physical functions and extractors when necessary.

Physical Function Infrastructure ($\mathcal{F}$)

The physical component p in a physical function pf is the result of a physical creation process $\mathsf{Create}_{a_{\mathsf{cr}}}$. The exact manifestation of a created physical component is determined by stochastic influences during the creation process, and by a controlled deterministic creation parameter a_{cr} which defines the full details of the creation process.

Definition 20 For a fixed tuple of parameters $(a_{\mathsf{cr}}, a_{\mathsf{ev}}, a_{\mathsf{ex}})$, the physical function infrastructure $\mathcal{F}_{(a_{\mathsf{cr}},a_{\mathsf{ev}},a_{\mathsf{ex}})}$ refers to the creation process $\mathsf{Create}_{a_{\mathsf{cr}}}$ and the set of all physical function systems consisting of the extraction algorithm $\mathsf{Extract}_{a_{\mathsf{ex}}}$, and a physical function $\mathsf{pf}_{\mathsf{p},a_{\mathsf{ev}}}$ with a physical component created by $\mathsf{Create}_{a_{\mathsf{cr}}}$:

$$\mathcal{F}_{(a_{\mathsf{cr}},a_{\mathsf{ev}},a_{\mathsf{ex}})} \stackrel{\Delta}{=} \left(\mathsf{Create}_{a_{\mathsf{cr}}}, \{\mathsf{pfs}_{\mathsf{p},a_{\mathsf{ev}},a_{\mathsf{ex}}} : \mathsf{p} \leftarrow \mathsf{Create}_{a_{\mathsf{cr}}}\}\right).$$

Finally, a family of physical function infrastructures is defined as a generalization of a physical function infrastructure for more than one single creation parameter a_{cr}:

$$\mathcal{F}_{(\mathcal{A}_{\mathsf{cr}},a_{\mathsf{ev}},a_{\mathsf{ex}})} \stackrel{\Delta}{=} \{\mathcal{F}_{a_{\mathsf{cr}},a_{\mathsf{ev}},a_{\mathsf{ex}}} : a_{\mathsf{cr}} \in \mathcal{A}_{\mathsf{cr}}\}.$$

If the parameters a_{ev} and a_{ex} are clear from the context, we simply write $\mathcal{F}_{a_{\mathsf{cr}}}$ and $\mathcal{F}_{\mathcal{A}_{\mathsf{cr}}}$. We use a family of physical function infrastructures to express the control one has over the creation procedure $\mathsf{Create}_{a_{\mathsf{cr}}}$, by being able to pick the creation parameter a_{cr} from a certain set $\mathcal{A}_{\mathsf{cr}}$.

Overview

In Fig. 3.2, we schematically show the concept of a physical function as the combination of a physical component and an evaluation procedure. Also shown is the expansion to a physical function system, by combining it with an extraction procedure, and further to a physical function infrastructure, by combining it with a creation procedure. Each of the three proposed procedures is possibly probabilistic and is for the remainder fully determined by its inputs and the respective parameters a_{ev}, a_{ex} and a_{cr}.

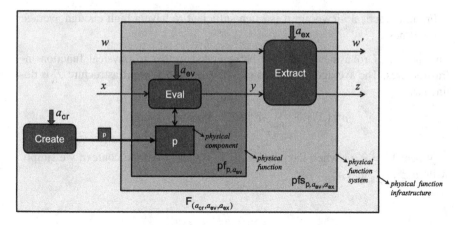

Fig. 3.2 Schematic overview of the concept of a physical function and its extension to a physical function system and a physical function infrastructure

3.3.4 Robustness of a Physical Function System

Robustness as defined in this section is the formal counterpart of reproducibility as described in Sect. 3.2.2, with the difference that reproducibility allows a certain error between evaluations as long as it is small, whereas robustness describes error-free reconstructions. In practice, an appropriate extraction algorithm is able to transform a reproducible PUF into a robust physical function system.

The robustness of a physical function system is defined as the probability that an output z produced during setup mode can later be reproduced exactly in reconstruction mode.

Definition 21 The challenge robustness of pfs with respect to $x \in \mathcal{X}$ is defined as

$$\rho_{\text{pfs}}(x) \stackrel{\triangle}{=} \Pr\big((z, w) \leftarrow \text{pfs}(x, w) : (z, w) \leftarrow \text{pfs}(x, \epsilon)\big).$$

When considering a subset of challenges $\mathcal{X}' \subseteq \mathcal{X}$, the following related robustness notions can be defined:

- Minimum robustness of pfs with respect to $\mathcal{X}'$:

$$\rho_{\text{pfs}}^{\min}(\mathcal{X}') \stackrel{\triangle}{=} \min_{x \in \mathcal{X}}\{\rho_{\text{pfs}}(x)\}.$$

This is useful when one requires every considered challenge to exhibit a minimal level of robustness.

- Average robustness of pfs with respect to $\mathcal{X}'$:

$$\rho_{\text{pfs}}^{\text{av}}(\mathcal{X}') \stackrel{\triangle}{=} \sum_{x \in \mathcal{X}} \rho_{\text{pfs}}(x) \cdot \Pr\big(x \leftarrow \mathcal{X}'\big).$$

From a practical viewpoint it is often sufficient to have a high enough average robustness.

This notion of robustness can even be extended further to physical function infrastructures. The average robustness of a physical function infrastructure $\mathcal{F}$ is defined as:

$$\rho_{\mathcal{F}}^{\mathrm{av}}(\mathcal{X}') \overset{\triangle}{=} \sum_{\mathsf{pfs} \in \mathcal{F}} \rho_{\mathcal{F}}^{\mathrm{av}}(\mathcal{X}') \cdot \mathrm{Pr}(\mathcal{F} \leftarrow \mathsf{Create}).$$

When $\mathcal{X}' = \mathcal{X}$ or when the content of $\mathcal{X}'$ is clear from the context, we simply write $\rho_{\mathsf{pfs}}^{\mathrm{min}}$, $\rho_{\mathsf{pfs}}^{\mathrm{av}}$ and $\rho_{\mathcal{F}}^{\mathrm{av}}$.

3.3.5 Physical Unclonability of a Physical Function System

Defining a (Physical) Clone

Before we define unclonability, we first need to agree on what we consider to be a clone. Intuitively, we consider two instances clones if they show 'similar behavior'. However, there are a number of technical details which have to be taken into account in order to turn this intuition into a formal definition.

First of all, we need to make explicit that we consider *physical* unclonability, and therefore also only *physical* clones. Given the definition of a physical function system, there are a number of non-physical ways one can think of to create 'similar instances'. For example, a physical function system deploying an extraction algorithm which outputs a fixed constant is trivial to clone, but it is obvious that we do not consider this a physical clone. A similar argument holds for the evaluation procedure of the physical function, e.g. imagine an evaluation procedure which quantizes a physical measurement into a zero bit response, which is basically a fixed value. We explicitly only consider physical clones by stating that two physical function systems which are considered clones can only differ in their physical component p, while their evaluation and extraction procedures and parameters need to be identical.

Secondly, there are many levels of 'similarity': two physical function systems can be 'less different than expected' or they can be truly 'indistinguishable'. This needs to be captured by a quantitative parameter. Also, we need to consider similarity with respect to a subset of challenges. If two physical function systems happen to coincide on a subset of challenges which is critical for a particular application, they need to be considered clones, even if they show completely different behavior on the remainder of challenges.

Finally, when attempting to formalize the notion of 'similarity' for physical function systems, we will run into robustness again. How does one define similarity with respect to a physical function system which even by itself does not always generate similar outputs? To tackle this issue, we are guided by two intuitive arguments: (i) every physical function system pfs should be a clone of itself (except for not deploying a different physical component), and (ii) a clone of a physical function system pfs cannot be more similar to pfs than pfs is to itself, as expressed by its robustness. In other words, the robustness of pfs is a natural upper bound for how similar a clone can be to pfs. Therefore, we express the similarity of a clone to pfs relative to its robustness.

Definition 22 For a fixed tuple of parameters $(a_{\mathsf{ev}}, a_{\mathsf{ex}})$, let $\mathsf{pfs}(= \mathsf{pfs}_{\mathsf{p}, a_{\mathsf{ev}}, a_{\mathsf{ex}}})$ and $\mathsf{pfs}'(= \mathsf{pfs}_{\mathsf{p}', a_{\mathsf{ev}}, a_{\mathsf{ex}}})$ be two physical function systems which are identical except for their physical components, $\mathsf{p} \neq \mathsf{p}'$. We say pfs is a δ-clone of pfs' with respect to $\mathcal{X}' \subseteq \mathcal{X}$, if $\forall x \in \mathcal{X}'$ it holds that

$$\Pr\big((z, w) \leftarrow \mathsf{pfs}'(x, w) : (z, w) \leftarrow \mathsf{pfs}(x, \epsilon)\big) \geq \delta \cdot \rho_{\mathsf{pfs}}(x),$$

with $0 \leq \delta \leq 1$. In shorthand notation, we write: $\mathsf{pfs} \overset{\delta; \mathcal{X}'}{\equiv} \mathsf{pfs}'$.

By $\mathsf{p} \neq \mathsf{p}'$ we mean that p and p' are distinct physical entities, i.e. occupying different positions in space-time, but they are allowed to be physically similar to any level of precision. Note that, except for their not deploying different physical components, every physical function system is a $(\delta = 1, \mathcal{X}' = \mathcal{X})$-clone of itself.

Defining Physical Unclonability

Now that we have a formal definition of a clone, we can define physical unclonability by formalizing the statement: 'it is difficult to produce a clone'. However, again some technicalities need to be considered before a formal description can be presented.

We first need to specify the capabilities of the adversary A that is trying to produce a clone. In practice, such an adversary will have access to a number of executions of the creation procedure of physical components. Possibly, it even has an amount of control over it, i.e. it can influence the physical processes taking place during creation within certain boundaries. We capture this formally by allowing the adversary to select the creation parameter a_{cr} from a particular subset $\mathcal{A}_{\mathsf{cr}}$. We use the notion of a family of physical function infrastructures $\mathcal{F}_{\mathcal{A}_{\mathsf{cr}}}$ to describe this. The adversarial model is described by means of a security game between the adversary and a creation oracle.

We also need to distinguish between two variants of unclonability:

- *Existential unclonability* means that it is hard to create a pair of physical function systems such that one is a clone of the other.
- *Selective unclonability* means that given a particular physical function system, it is hard to create a second one which is a clone of the first one.

Note that in general, existential unclonability implies selective unclonability and is therefore a stronger security notion. We will give a formal definition for existential unclonability, but the same approach as presented here can be applied to describe selective unclonability.

To formally define (existential) unclonability (or cloning resistance), we first describe the adversary model A by means of a *cloning game* $\mathbf{Game}_A^{clone}(\mathcal{A}_{cr}, q)$:

- In the cloning game $\mathbf{Game}_A^{clone}(\mathcal{A}_{cr}, q)$, an adversary is allowed to make up to q queries to a creation oracle $O_{\mathcal{A}_{cr}}^{Create}$, with $q \geq 2$.
- The creation oracle $O_{\mathcal{A}_{cr}}^{Create}$ expects a creation parameter a_{cr} as query input. If $a_{cr} \in \mathcal{A}_{cr}$, then the oracle invokes the physical component creation procedure with the queried parameter to create a single physical component $\mathsf{Create}_{a_{cr}} \to \mathsf{p}$ and answers the query with p.
- The adversary is allowed to adaptively change the creation parameter a_{cr} of its queries.
- When the game ends, the adversary is required to output a pair of physical components $(\mathsf{p}, \mathsf{p}')$ both of which it received as a query reply from the creation oracle during the game.

Definition 23 A family of physical function instantiations $\mathcal{F}_{(\mathcal{A}_{cr}, a_{ev}, a_{ex})}$ is (γ, δ, q)-cloning-resistant with respect to $\mathcal{X}' \subseteq \mathcal{X}$ if for every probabilistic polynomial time adversary A it holds that:

$$\Pr\left(\mathsf{pfs}_{\mathsf{p}, a_{ev}, a_{ex}} \overset{\delta; \mathcal{X}'}{\equiv} \mathsf{pfs}'_{\mathsf{p}', a_{ev}, a_{ex}} : (\mathsf{p}, \mathsf{p}') \leftarrow \mathbf{Game}_A^{clone}(\mathcal{A}_{cr}, q)\right) \leq \gamma.$$

The level of control an adversary has over the creation process will to a large extent determine the cloning resistance of a physical function infrastructure family. As explained, the influence an adversary has over the creation is represented by $\mathcal{A}_{cr}$. We distinguish a special case when $\mathcal{A}_{cr} = \{a_{cr}\}$, i.e. the creation process is fixed. This is typically so for the genuine manufacturer of the physical function systems and therefore we call this the *honest manufacturer* adversary model. Note that even the honest manufacturer can coincidentally create clones but this should only happen with low probability. In all other cases with more than one element in $\mathcal{A}_{cr}$, it means that the manufacturer is deliberately influencing the creation process in order to create a clone. This is called the *malicious manufacturer* adversary model.

3.3.6 Unpredictability of a Physical Function System

When using a physical function system pfs in a security application, the unpredictability of its output is the most basic expected security requirement. In classic cryptography, unpredictability is a well-established concept, e.g. for random functions, and it expresses the difficulty of predicting an unobserved output of a function after having observed different function evaluations. Due to the peculiarities of physical function systems, as discussed in detail in the previous paragraphs, the classical definition of unpredictability does not directly apply. We will adapt it in an appropriate manner as to make it capture the notion of unpredictability for physical function systems.

Types of Unpredictability

We distinguish between two different types of unpredictability for physical function systems: unpredictability with respect to different outputs on the same system, and unpredictability with respect to the same outputs on different systems. The first type is typically important when the physical function system is used as a challenge-response entity, e.g. in an authentication protocol. It would be highly undesirable if the response in the next run of the protocol could be predicted based on previously observed runs of the protocol. The second type is mainly of significance when the physical function system is used as a secure storage mechanism, e.g. to generate cryptographic keys. In that case, the independence of outputs from different physical function systems is of the utmost importance to ensure the randomness of the derived keys. Note that the first type of unpredictability is a direct extension of the classical unpredictability notion to physical function systems, and a theoretical definition will be a formal variant of the unpredictability property of PUFs as discussed in Sect. 3.2.5. The second type of unpredictability, on the other hand, could be regarded as a generalization of the uniqueness property of PUFs as discussed in Sect. 3.2.3. Whereas uniqueness only requires that different PUF instances produce sufficiently different responses, this type of unpredictability additionally requires a more stringent apparent independence between the outputs of different physical function systems. The formalization we propose next captures both types of unpredictability in a single definition, as well as the continuum of intermediate cases.

Defining Unpredictability for Physical Function Systems

We again use a game-based approach to describe a model for the adversary A. We distinguish between a weak and a strong prediction game, based on the control the adversary has over picking the physical function systems and challenges which are evaluated.

- The *weak prediction game* $\mathbf{Game}_A^{predict;weak}(\mathcal{P}_{learn}, \mathcal{P}_{chal}, q)$ is a game between an adversary A and an evaluation oracle $O_{\mathcal{P}_{learn}, \mathcal{P}_{chal}}^{Eval}$ which takes place in two phases: a learning phase and a challenge phase.
- During the learning phase, A is allowed to query the oracle up to q times. When queried, $O_{\mathcal{P}_{learn}, \mathcal{P}_{chal}}^{Eval}$ randomly selects $pfs_i \xleftarrow{\$} \mathcal{P}_{learn}$ and $x_i \xleftarrow{\$} \mathcal{X}$ and evaluates $(z_i, w_i) \leftarrow pfs_i(x_i, \epsilon)$. For each of these q queries, the adversary A is allowed to observe the tuple (pfs_i, x_i, z_i, w_i).
- During the challenge phase, $O_{\mathcal{P}_{learn}, \mathcal{P}_{chal}}^{Eval}$ randomly selects $pfs \xleftarrow{\$} \mathcal{P}_{chal}$ and $x \xleftarrow{\$} \mathcal{X}$, making sure that the combination (pfs, x) was never evaluated during

the learning phase, and evaluates $(z, w) \leftarrow \mathsf{pfs}(x, \epsilon)$. The adversary A is now allowed to observe the tuple (pfs, x, w).

- At the end of the game, the oracle outputs the tuple (pfs, x, z) and the adversary outputs a prediction z'.

The *strong prediction game* $\mathbf{Game}_{\mathsf{A}}^{\text{predict; strong}}(\mathcal{P}_{\text{learn}}, \mathcal{P}_{\text{chal}}, q)$ is completely equivalent to the weak prediction game, only now physical function systems and challenges are no longer randomly selected by the oracle, but are adaptively queried by the adversary. In the learning phase, the adversary queries the oracle with up to q tuples $(\mathsf{pfs}_i \in \mathcal{P}_{\text{learn}}, x_i, w_i)$ and learns the full evaluations $(z_i, w'_i) \leftarrow \mathsf{pfs}_i(x_i, w_i)$ from the oracle. In the challenge phase, the adversary queries the oracle with a single tuple $(\mathsf{pfs} \in \mathcal{P}_{\text{chal}}, x, w)$ such that the combination (pfs, x) was never queried during the learning phase. The oracle evaluates $(z, w') \leftarrow \mathsf{pfs}(x, w)$ but A can only observe w' this time.

Definition 24 Let $\mathcal{P}_{\text{learn}}$ and $\mathcal{P}_{\text{chal}}$ be subsets containing physical function systems from the same physical function infrastructure $\mathcal{F}$. We say the physical function systems in $\mathcal{P}_{\text{chal}}$ are (λ, q)-weakly unpredictable with respect to $\mathcal{P}_{\text{learn}}$ if for every probabilistic polynomial time adversary A, it holds that:

$$\Pr\left(z = z' : \big((\mathsf{pfs}, x, z), z'\big) \leftarrow \mathbf{Game}_{\mathsf{A}}^{\text{predict; weak}}(\mathcal{P}_{\text{learn}}, \mathcal{P}_{\text{chal}}, q)\right) \leq \lambda \cdot \rho_{\mathsf{pfs}}(x).$$

Similarly, we say the physical function systems in $\mathcal{P}_{\text{chal}}$ are (λ, q)-strongly unpredictable with respect to $\mathcal{P}_{\text{learn}}$ if for every probabilistic polynomial time adversary A, it holds that:

$$\Pr\left(z = z' : \big((\mathsf{pfs}, x, z), z'\big) \leftarrow \mathbf{Game}_{\mathsf{A}}^{\text{predict; strong}}(\mathcal{P}_{\text{learn}}, \mathcal{P}_{\text{chal}}, q)\right) \leq \lambda \cdot \rho_{\mathsf{pfs}}(x).$$

Note that $\mathcal{P}_{\text{chal}}$ and $\mathcal{P}_{\text{learn}}$ do not need to be mutually exclusive and can even be identical. In fact, if $\mathcal{P}_{\text{chal}} = \mathcal{P}_{\text{learn}} = \{\mathsf{pfs}\}$, i.e. we only consider the unpredictability of a single pfs with respect to itself, then Definition 24 closely resembles the classical notion of unpredictability of a random function.

3.3.7 Discussion

Most of the concepts and properties defined in this section are more rigid formalizations of concepts and properties which we introduced earlier in a more intuitive way, respectively in Sects. 2.2.1 and 3.2.

- The formally defined notion of a physical function infrastructure $(\mathcal{F})$ coincides almost completely with what we, rather intuitively, have called a PUF class $(\mathcal{P})$ in Sect. 2.2.1. Both describe a set of instantiations and a creation procedure (Create).
- A PUF instance (puf), described in Sect. 2.2.1 as having a *physical state* which can be measured by an evaluation procedure (Eval), is equivalent to a formal physical function (pf) consisting of a *physical component* (p) and the same evaluation procedure.

- Concrete extractor constructions have not yet been described, but are treated in detail in Chap. 6. It is evident that they are captured formally by the rather generic extraction procedure introduced in this section. The formal concept of a physical function system coincides with the concatenation of a PUF with an extractor implementation.
- The formally defined property of robustness of a physical function system is an extension of the earlier introduced PUF property of reproducibility (cf. Definition 7), taking into account the effect of the extractor.
- Equivalently, the formal definition of cloning resistance is the extension of physical unclonability (cf. Definition 10), related to the formal version of robustness.
- The formal notion of unpredictability is defined in a broad sense, considering both unpredictability with respect to the same instance as well as regarding other instances. In that aspect, the formal unpredictability is to be considered as the formalization of the earlier introduced unpredictability notion (cf. Definition 11, describing unpredictability with regard to different responses on the same PUF instance) in combination with a generalized version of uniqueness (cf. Definition 8, describing *differentness* of responses with regard to other PUF instances).

The resemblance between these formal definitions and the rather intuitive concepts and properties which we have defined earlier makes clear that the proposed framework is of a significant practical value. This is strengthened by the fact that most of the intuitive properties have been experimentally verified for many PUF implementations and can hence be directly translated to their formal counterparts. Evidently, to do this, the effect of the extractor needs to be taken into account.

3.4 Conclusion

Following an extensive study of 11 meaningful PUF properties on a representative subset of PUF constructions, and on a reference set of non-PUF constructions, we are able to extract the defining properties of a PUF: *identifiability and physical unclonability*, and by implication constructibility, evaluability, reproducibility and uniqueness. The remaining properties of unpredictability, mathematical and true unclonability, one-wayness and tamper evidence are classified as *nice-to-have*, but are not strictly required for a construction to be called a PUF. This also means that many PUF proposals in fact do not exhibit these nice-to-have properties, or at the very best it remains as yet unclear whether they do. In this light, the optical PUF as proposed by Pappu et al. [105] serves as an exemplary prototype PUF, meeting all mentioned properties to a major extent. Most intrinsic PUFs fall short on the property of mathematical unclonability, and the existence of a *strong intrinsic PUF* with motivated and verified security guarantees remains an open question. One-wayness turns out to be a particularly unfitting property for intrinsic PUFs, which is partially caused by the ambiguity of considering the one-wayness of PUFs. Tamper evidence, while often proclaimed as one of the major advantages of using PUFs, remains a question mark for all intrinsic PUF proposals, as no experimental results on any construction are known.

Based on these results and with the aim of formalizing these particularly interesting properties, we introduce a framework for working with PUFs (and physical functions in general) in a theoretical security setting. Using the introduced primitives in this framework, we formally define robustness, physical unclonability and unpredictability of a physical function.

Chapter 4
Implementation and Experimental Analysis of Intrinsic PUFs

4.1 Introduction

4.1.1 Motivation

All currently known intrinsic PUF implementations are silicon-based and obtain their PUF behavior from process variations during the manufacturing of silicon chips. These PUFs are of particular interest, because their response values can be used as a secret element in a larger security implementation on the same silicon chip. Deploying an intrinsic PUF in a security application in this way provides interesting practical and security advantages. We will discuss applications of intrinsic PUFs and their added value in great detail, respectively for PUF-based authentication in Chap. 5 and PUF-based key generation in Chap. 6.

The realization that a silicon intrinsic PUF can serve as an integral hardware security primitive with valuable properties such as uniqueness and unpredictability, and in particular physical unclonability, which cannot be obtained solely from algorithmic constructions, has led to a great interest into its constructions. Many intrinsic PUF implementations were proposed over time and are discussed in detail in Sect. 2.4. When one actually wants to deploy an intrinsic PUF in a hardware security system, interest goes out to the levels of efficiency and performance of all these different constructions, both from a PUF perspective (which construction shows the best PUF behavior) as well as from a typical hardware design perspective (which construction offers the lowest area, highest speed, lowest power use, etc.). Section 2.4.8 provides an overview of known experimental results concerning PUF behavior, but it is also made clear that a comparison of different proposals solely based on these results is not entirely objective, for a number of reasons:

1. For all known intrinsic PUF constructions, as for nearly all hardware security primitives, there exists a trade-off between area/speed and security, i.e. larger and/or slower implementations typically offer higher levels of security. In the overview in Sect. 2.4.8, which lists experimental PUF results available in literature, some of the considered implementations are heavily biased towards opti-

R. Maes, *Physically Unclonable Functions*, DOI 10.1007/978-3-642-41395-7_4,
© Springer-Verlag Berlin Heidelberg 2013

mizing their PUF behavior by greatly sacrificing on implementation area and/or speed, while others are not.

2. Experimental PUF results always focus on two properties, reliability and uniqueness, which are mostly summarized by calculating the average response intra-distance ($\mu_\mathcal{P}^{\text{intra}}$) and inter-distance ($\mu_\mathcal{P}^{\text{inter}}$) of the experiment. Good PUF behavior is expressed by a small $\mu_\mathcal{P}^{\text{intra}}$ and a large $\mu_\mathcal{P}^{\text{inter}}$. However, it is not immediately clear how to combine both measures into a single quality parameter for a particular PUF construction.

3. The results presented in Sect. 2.4.8 come from implementations using a variety of different technologies and platforms. It is typically hard to accurately scale implementation results such as area and speed from one technology to another, e.g. between different CMOS technology nodes, and scaling between different platforms, e.g. from FPGA to ASIC, can only be done very roughly. Extrapolation of PUF behavior results to different technologies and platforms is virtually impossible.

The first two issues cannot be dealt with for bare PUFs, but need to be considered in the application context of the bigger system deploying the PUF. The optimal trade-off between security and efficiency, as well as the relation between reliability and uniqueness, are determined by the requirements and constraints of the security system as a whole. Optimizing the PUF implementation is an entangled part of a bigger design optimization process which will be discussed in detail for PUF-based authentication systems in Chap. 5 and for PUF-based cryptographic key generation in Chap. 6.

The last issue can be approached by implementing different intrinsic PUF proposals on the same platform and using the same technology. This will be the main topic of this chapter. A selection is made of intrinsic PUF proposals which were proven to show acceptable PUF behavior, and a number of instantiations of each of them is integrated in an ASIC design. This ASIC design is processed and a significant set of silicon chips implementing it is manufactured. Based on this set of devices, valuable experimental PUF data is gathered which can be cross-compared without restraint, since it results from PUF constructions implemented on the same silicon die.

Background   Designing and manufacturing an ASIC is a complex, time-consuming, and costly undertaking with significant risk of failure. The findings presented in this chapter are the joint successful result of a European research project called 'UNIQUE' in which the author's institution was a partner [149]. We definitely want to acknowledge the other project partners which contributed heavily to the design, production and evaluation of the test chip discussed in this chapter.

4.1.2 Chapter Goals

The main goal of this chapter is to produce a practical and an objective analysis and comparison of different intrinsic PUF constructions, by implementing them on the same platform, evaluating them under the same conditions, and assessing them on the same characteristics. In this chapter we plan to:

- Discuss the design process and implementation details of a custom test chip carrying instantiations of six different intrinsic PUFs.
- Present an in-depth analysis of the uniqueness and reproducibility of the evaluation results of the test chip implementations, including the influence of varying evaluation conditions. This analysis should result in a practically usable characterization of these properties for every PUF instance, which can be immediately plugged into the design and optimization process of an application seeking to deploy one of these PUF implementations.
- Discuss the notion of PUF response entropy and introduce a number of practical entropy bounds which can be computed based on experimentally obtained response evaluations, including a method to calculate a PUF response entropy bound based on the results of a modeling attack.

4.1.3 Chapter Overview

Section 4.2 discusses the realization of the intrinsic PUF test chip, from its initial design rationale and requirements, through its architecture and the description of its building blocks, to its manufacturing flow details. In Sect. 4.3, we describe how the test chip samples were evaluated and we present a complete analysis of uniqueness and reproducibility based on a large data set of evaluation results. The entropy of a PUF response is described in Sect. 4.4 as a measure of its unpredictability. We introduce a number of increasingly tighter bounds on the response entropy of a PUF, based on considering increasingly more advanced adversary models. Finally, the main results of this chapter are summarized in Sect. 4.5.

4.2 Test Chip Design

4.2.1 Design Rationale

The rationale behind the design of the test chip is guided by two main considerations:

(i) In the end, the goal of the test chip is to collect statistically significant experimental data from intrinsic PUF implementations. Ideally, we would like to implement as many and as large instances as possible from as many different intrinsic PUF proposals as possible and evaluate them in a quick, easy and realistic manner, taking into account that the available area budget should be more or less evenly distributed among the different PUFs.

(ii) Designing and producing an ASIC is a very complex process with a minimal margin for error. The probability of critical failures increases steadily with the size and complexity of the design. Since the coordinating project provides only a single opportunity for ASIC production, it needs to be first-time-right and any risk of failure should be minimized.

These two considerations lead to the following design choices:

- To minimize risk, the overall architecture is kept minimalistic, with the major portion of the silicon area budget devoted to implementations of PUF instances.
- We mainly choose to implement PUF constructions which (at the time) were proven to show PUF behavior in earlier experiments.
- The additional components are kept to the bare minimum required to realistically evaluate the PUFs. This leaves the most area to the actual PUF implementations. All measurement post-processing is done off-line.
- The measurement communication interface, being a single-point-of-failure in the whole design, is kept as simple as possible to minimize all risk. This comes at a significant sacrifice in measurement speed.
- Whenever possible, standard design flows are used. Except for the low-level implementation of some of the PUFs, the whole design is described at the RTL level and synthesized using reliable third-party standard cell libraries for the considered technology.

4.2.2 Design Requirements

PUF Selection

Six intrinsic PUF structures are selected for integration on the test chip:

1. The ring oscillator PUF as proposed by Suh and Devadas [136].
2. The latch PUF as proposed by Su et al. [135].
3. The SRAM PUF as proposed by Guajardo et al. [45].
4. The D Flip-Flop PUF as proposed by Maes et al. [84].
5. The arbiter PUF as proposed by Lee et al. [75].
6. The buskeeper PUF as proposed by Simons et al. [132].

The first five PUF constructions are selected because they were proven to show PUF behavior in earlier implementations. The buskeeper PUF is a newly proposed PUF construction by Simons et al. [132]. For the ring oscillator and the arbiter PUF structures, we analyse two different evaluation methods (cf. Sect. 4.3.1) resulting in a total of eight different PUF constructions on the chip.

Evaluation Control

Additional control over the evaluation conditions of the selected PUF implementations is desirable:

- In order to study the effects of the power-up conditions on the PUF constructions which depend on power-up behavior, a number of PUF instances are grouped under a separate power supply on the chip. This allows us to test instances on the same chip under different supply voltage conditions.
- In a realistic application, a PUF is integrated on the same silicon die implementing the complete hardware system, including a large amount of rapidly switching logic. This might have an effect on the PUF's behavior, e.g. because it introduces switching noise on the supply voltage. To mimic this behavior, we implement an active core on the test chip whose sole purpose is to generate switching activity while the PUFs are evaluated.

Interfacing

To transfer the measurement data off-chip we require a communication interface which offers reasonable transfer rates at minimal design complexity and low pin-count. We prefer a standardized interface for easy integration with other components. A Serial Peripheral Interface (SPI) [98] was selected.

Internally, the different building blocks on the chip need to be accessible in a straightforward manner. However, since all blocks operate in slave-mode, we don't require advanced communication control and we don't want to dedicate silicon area to complex bus interfaces. We opted for a memory-mapped organization, with a single address decoder controlling which building block is being read from or written to. Since most selected PUF blocks are inherently memory elements, they are trivially integrated in this organization. For the other building blocks, input and output ports are being accessed through addressable registers.

4.2.3 Top-Level Architecture

Figure 4.1 indicates the top-level architecture of the test chip design. All data communication is done through the SPI/memory-mapped interface, except for the basic clock and power domain controls which has dedicated control and status pins. The SPI encoder and decoder is a standard design supporting the SPI protocol. The active core is basically an implementation of a large number of unrolled rounds of a random substitution/permutation layer, with the only intention of generating a lot of switching activity. The memory mapper consists of a large multiplexer and demultiplexer which direct data to and from the addressed building block. Internally, data interfaces are 32-bit signals and the address is a 19-bit signal.

Figure 4.2 details the memory map of the different building blocks onto the address space. The three most significant address bits are used to select a particular building block. The next four address bits select a particular instance within the building block, e.g. a particular instance of a PUF type. The remaining 12 address bits are used for addressing within a single instance.

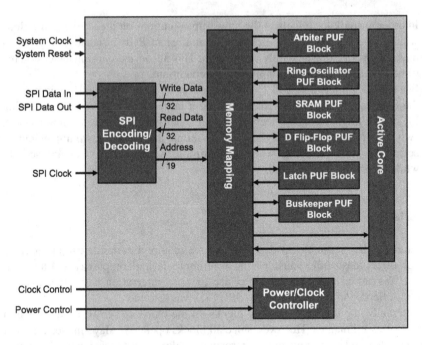

Fig. 4.1 Top-level block diagram of the test chip

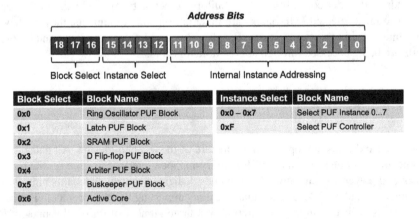

Fig. 4.2 Address structure of the internal memory map of the test chip

4.2.4 PUF Block: Arbiter PUF

We design the arbiter PUF according to the original construction from Lee et al. [75], as shown in Fig. 2.1. The switch blocks are constructed using two 2-to-1 multiplexers. The arbiter is an SR latch consisting of two cross-coupled NAND-gates. Each arbiter PUF has a delay chain consisting of 64 concatenated switch blocks, which is long enough to accumulate sufficient delay randomness in order to exhibit an observable difference between the two lines and with high probability avoid the arbiter going into the metastable state. This also means the arbiter PUF takes 64-bit challenges. Using more switch blocks gives even longer challenges, but does not substantially increase the arbiter PUF's unpredictability and does result in larger area use and slower evaluation. To minimize bias in the delay lines and in the arbiter circuit, the whole arbiter PUF is a full custom design, i.e. all design steps including the geometrical sizing, placement and routing of the transistors and interconnecting metal lines are done by hand. Special attention is paid to the symmetry of the arbiter circuit and to balancing the parasitic capacitances of the delay lines as closely as possible. The test chip contains 256 instantiations of this arbiter PUF design, which are grouped into eight instances of 32 arbiter PUFs each. This grouping is only for evaluation performance reasons (32 arbiter PUF response bits can be read out simultaneously over the 32-bit data bus), since all instantiations are identical.

4.2.5 PUF Block: Ring Oscillator PUF

The design of the ring oscillator PUF is based on the construction from Suh and Devadas [136] which is depicted in Fig. 2.3. A ring oscillator consists of 80 chained inverters and one NAND gate to control the oscillation. The number of looped inverters roughly determines the nominal frequency of the ring oscillator, and in the order of 60~80 inverters are required to obtain frequencies in the range of 500~700 MHz, which are countable with regular digital counters in the targeted technology. Of this ring oscillator, 4096 identical copies are implemented on the test chip, arranged in 16 batches of 256 oscillators each. Every batch has a single frequency counter and a 256-to-1 multiplexer connects one of the batch's oscillators to the counter. To cope with the high frequency oscillations, the counter is implemented as a 32-bit toggle counter which has a very short critical path. Since there are 16 batches, each with its own counter, 16 oscillation frequencies can be measured in parallel. The measurement time during which oscillations are counted is determined as a particular number of oscillations of an independent, slightly faster, oscillator (64 inverters + NAND) which feeds a timer. The exact number of oscillations of this timing loop can be set by the user. The actual response bit generation is not implemented on the test chip, but the counter values are measured directly. The response bit generation based on the measured frequencies is performed off-line, using an algorithm of one's choice. The response evaluation methods we use are detailed in Sect. 4.3.1.

4.2.6 PUF Block: SRAM PUF

The SRAM PUF, as proposed by Guajardo et al. [45], basically consists of standard SRAM cells of which the power-up value is measured. We implement an SRAM

PUF using a third-party (TSMC) SRAM IP block implementing an addressable array of 2048×32 SRAM cells, each cell consisting of six MOSFETs. Each of these blocks can generate 65536 response bits (64 kbit). Four of these SRAM PUF instances are placed on the test chip.

4.2.7 PUF Blocks: D Flip-Flop PUF, Latch PUF and Buskeeper PUF

The design of these three PUFs basically consists of instantiations of the basic elements: D flip-flops, latches and buskeeper cells. For our test chip, the operation of all three of these PUFs is based on the power-up behavior of their basic cells. The design of each of these cells comes from a third-party (TSMC) standard cell library. The only other difference between these PUFs is the number of cells which are instantiated and the way the cells are organized in arrays.

D Flip-Flop PUF Block

One D flip-flop PUF is designed containing 8192 D flip-flop standard cells. Four of these PUFs are instantiated on the test chip. In the first two instances, the D flip-flops are organized in a long scan chain, allowing them to be read out sequentially. In the last two instances, the flip-flops are organized in a large multiplexer tree, allowing them to be addressed individually. The flip-flops in the multiplexer tree have their data inputs connected to the write input of the memory map, which is grounded when the D flip-flop PUF is not addressed.

Latch PUF

We design a latch PUF which consists of 8192 standard cell latches, and four of these latch PUFs instances are implemented on the test chip. Again, the first two instances have a scan chain-based organization while the latter are organized in a multiplexer tree. For latches, the scan chain design is slightly more advanced. Because latches are level-triggered, as opposed to flip-flops which are edge-triggered, they cannot be all clocked at the same time to shift their values in a chain. The latches in the multiplexer tree have their data inputs grounded.

Buskeeper PUF

The buskeeper PUF design contains 8192 buskeeper cells, and two of the buskeeper PUFs are instantiated on the test chip. Both are organized in an addressable multiplexer tree, since buskeeper cells cannot be chained (they have only a single-bit bidirectional port).

Practical Comparison

D flip-flop, latch and buskeeper PUFs are very similar in design, the main difference being the implementation of their basic cells. The following practical considerations can be made:

- Buskeeper cells, consisting of two inverters, are the smallest of the three, which makes the buskeeper PUF the most area-efficient in terms of response bits per silicon area. However, since they cannot be chained, they do require a (rather large) multiplexer tree to read them out.
- Latches, which consist of two cross-coupled NAND or NOR gates, are larger than buskeeper cells, but smaller than D flip-flops. They can be chained, but this is not trivial.
- D flip-flops are typically constructed from two latches and are therefore the largest of all three basic cells, but they can be easily chained. Moreover, D flip-flops are also a very common cell in regular digital designs, which could make them reusable for other purposes after they have generated their PUF response bit at power-up.

In comparison to SRAM PUFs, these three PUF types are less efficient since SRAM arrays are heavily area-optimized. However, their cell-based design allows us more flexibility since they can be instantiated one cell at a time, while SRAM only comes in bulky arrays. This also allows us to spread, e.g. a D flip-flop PUF, randomly over the whole area of a silicon die, which adds a layer of physical obscurity against optical scrutiny attacks. SRAM arrays on the other hand are easily spotted due to their large and very regular matrix structure.

Besides these practical observations, the PUF behavior of each of these memory-based PUFs should of course also be taken into account. This is the goal of the test chip as discussed in this chapter.

4.2.8 Power Domains

The test chip design contains two separate core power domains. The primary goal of the separated power domain is to ease reading out the memory-based PUFs which require a power cycle. This way, the main part of the test chip core containing the communication interface can stay active while some of the memory-based PUFs in the separate power domain are power-cycled to generate a new response. Besides this goal, the separate power domain is also convenient when performing reliability tests under varying supply voltages.

The separate power domain contains one SRAM PUF instance (out of four), one D flip-flop PUF instance (out of four), one latch PUF instance (out of four), and one buskeeper PUF instance (out of two). The separate power domain has independent supply voltage and ground pins. Moreover, all signal lines connecting the two power domains can be blocked, effectively electrically isolating the PUF instances in the separate power domain.

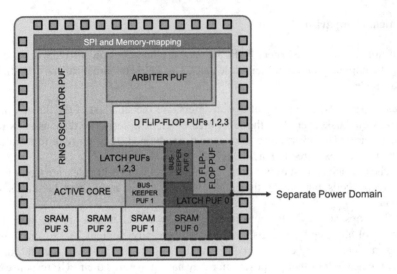

Fig. 4.3 Floor plan of the structures on the test chip

4.2.9 Implementation Details

Floor Plan

The schematic floor planning for the different PUF instances and other building blocks on the test chip's silicon die area is shown in Fig. 4.3. The separate power domain is also depicted, and as shown it contains one instance from all four memory-based PUFs. The active core is placed in the free space in between the different instances, in order to increase the impact of its toggling activity on the PUF evaluations.

Development Flow

Except for the arbiter and ring oscillator PUF blocks, the whole test chip design is described at the RTL level using a hardware description language (VHDL) and synthesized using a standard cell library. The back-end design is done by an external party (Invomec). The arbiter PUF is designed as a full-custom layout to have the most control over delay line and arbiter circuit balancing. The ring oscillator PUF is designed as an array of identically laid out hard-macro copies of an inverter chain. This is to make sure that all oscillators have the same nominal frequency, and any frequency difference is only caused by silicon process variations.

Implementation Technology

The final design of the test chip is implemented in 65 nm low-power CMOS technology (TSMC 65 nm CMOS Low Power MS/RF and TCBN65LP (nominal Vt)

Table 4.1 Silicon area breakdown of the different test chip building blocks

Building block	Silicon area (mm^2)	Relative area (·/total logic)	Building block content
Ring Oscillator PUF	0.241	10.7 %	4096 ring oscillators + 16 × 32-bit counters + control
Latch PUF	0.272	9.5 %	4 × 8192 latches + 2 × multiplexer tree
SRAM PUF	0.213	12.1 %	4 × 64 kbit SRAM array
D Flip-Flop PUF	0.392	17.4 %	4 × 8192 D flip-flops + 2 × multiplexer tree
Arbiter PUF	0.279	12.4 %	256 × 64-bit arbiter PUF + control
Buskeeper PUF	0.076	3.4 %	2 × 8192 buskeeper cells + 2 × multiplexer tree
Active Core	0.353	15.7 %	32 × 128-bit substitution-permutation rounds
Additional Blocks	0.425	18.9 %	SPI interface, memory mapping, power control, …
Total Logic Area	2.251	100.0 %	all of the above
Overhead	1.405	62.4 %	I/O pads, power/ground rings, empty space, …
Complete Test Chip	3.656	162.4 %	1912 μm × 1912 μm silicon die

standard cell library). Both power domains of the core logic are nominally powered by $V_{dd} = 1.2$ V, and the I/O voltage is $V_{io} = 2.5$ V. The resulting silicon die is packaged in an LQFP64 package. In total, 192 packaged chips are produced.

Area Breakdown

Table 4.1 shows an estimate of the silicon die area breakdown of the different building blocks on the test chip. The estimates in this table are used in Chaps. 5 and 6 to estimate the required PUF size for a PUF-based application with given requirements. This provides an as objective as possible comparison between the different PUF constructions.

4.3   Experimental Uniqueness and Reproducibility Results

4.3.1   Evaluation of Delay-Based PUFs

Before we present the statistics of the experimental data, we first need to describe the manner in which we evaluate bit responses for the delay-based PUF constructions, i.e. the arbiter and the ring oscillator PUFs. The evaluation of the memory-based PUFs follows trivially from their design.

Arbiter PUF Evaluation Modes

The basic arbiter PUF already produces single bit responses. We also consider 2-XOR arbiter PUFs as proposed by Majzoobi et al. [94] by pairing up arbiter PUFs and perform an XOR-operation on their outputs to produce a single bit response. The XOR-operation is not implemented on the test chip but is performed off-line on the evaluated response bits of the basic arbiter PUFs.

Ring Oscillator PUF Evaluation Modes

As mentioned in Sect. 4.2.5, the ring oscillator PUF design outputs the frequency counter values directly. While these values already show some PUF behavior, it is difficult to use them as such in an application. For ease of integration, the frequency counter values need to be encoded in a meaningful binary response format. This will also make the comparison with the other PUF types more relevant. Ring oscillator PUF response bits are typically generated based on the relative comparisons between measured frequencies, as these comparisons are much more resilient to noise and varying evaluation conditions than the absolute frequency values. We present two encoding methods based on relative orderings of the frequency counter values.

The first method is a basic pairwise comparison (P.C.) between counted frequencies from different but simultaneously measured oscillators, as was proposed by Suh and Devadas [136] (but without the 1-out-of-k masking technique). A single response bit is generated based on the outcome of each comparison. To ensure independent responses, every oscillator is only used for the generation of a single bit. The arrangement of our ring oscillator PUF design in 16 batches, with 256 oscillators and one frequency counter per batch, allows us to measure 16 frequencies simultaneously and hence produce eight response bits in one evaluation. Using this evaluation method, the complete ring oscillator PUF design can generate $256 \times 8 = 2048$ response bits.

In [87], we developed a new response bit generation method for ring oscillator PUFs based on the ordering of the measured frequencies. Suh and Devadas [136] and Yin and Qu [157] observed that the amount of information in the ordering of n independent and identically distributed frequencies is as high as $\log_2 n!$. If one can find an efficient and noise-resilient encoding of such an ordering, this will lead to a significant increase in the number of independent response bits, since $\log_2 n! = \sum_{i=2}^{n} \log_2 i \approx n \cdot \log_2 \frac{n}{e}$ is superlinear in n, as opposed to the pairwise comparison method, which can only produce a number of response bits that is linear in n, i.e. $\frac{n}{2}$.

In [87], we propose using a Lehmer encoding [76, 125] to represent the ascending order of a vector of simultaneously measured frequencies, followed by a Gray encoding [44] of the Lehmer coefficients. A Lehmer code is a unique numerical representation of an ordering (permutation) which is moreover efficient to obtain since it does not require explicit value sorting. If $f^n = (f_1, \ldots, f_n)$ is a vector containing n frequency measurements, then the Lehmer code of the ascending order of these values is a coefficient vector $r^{n-1} = (r_1, \ldots, r_{n-1})$ with $r_i \in \{0, 1, \ldots, i\}$. It is clear that r^{n-1} can take $2 \times 3 \times \cdots \times n = n!$ possible values, which is exactly the number of possible orderings of f^n; hence each ordering has a unique Lehmer code representation. The Lehmer coefficients are calculated from f^n as $r_j = \sum_{i=1}^{j} gt(f_{j+1}, f_i)$, with $gt(x, y) = 1$ if $x > y$ and 0 otherwise. The Lehmer encoding has the nice property that a minimal change in the sorted ordering, caused by two neighboring values swapping places after sorting, only changes a single Lehmer coefficient by ± 1. Using a binary Gray encoding for the Lehmer coefficients, this translates to only a single bit difference, which makes the overall response bit generation particularly noise-resilient. The length of the binary representation becomes $\sum_{i=2}^{n} \lceil \log_2 i \rceil$, which is a nearly optimal representation of the actual amount of information in the ordering, i.e. $\sum_{i=2}^{n} \log_2 i$.

For our ring oscillator PUF design on the test chip, we apply this Lehmer-Gray (L.G.) encoding off-line on each vector of $n = 16$ simultaneously measured frequencies, yielding 49 response bits. In total, the ring oscillator PUF can generate $256 \times 49 = 12544$ response bits using this method, which is over six times more than when using the pairwise comparison method.

4.3.2 PUF Experiment: Goals, Strategy and Setup

Experiment Goals

The goal of the experiments on the test chip is to characterize the meaningful properties of the eight studied PUF constructions as realistically and as accurately as possible. As explained in Sect. 4.1.1, it is not possible to make a comprehensive ranking of the different PUFs solely based on their bare characteristics. In order to make such an objective comparison, the envisioned application needs to be taken into account as well. We will do this respectively in Chaps. 5 and 6 for PUF-based authentication and PUF-based key generation. However, a common interface is desirable, i.e. a uniform set of characterization parameters for the meaningful properties of every PUF construction, which can be used in an unambiguous manner to determine their usability in a particular application. Here, we will provide such characterization parameters for the uniqueness and reproducibility of all studied PUF constructions, respectively in Sects. 4.3.3 and 4.3.4. The unpredictability of the different PUF constructions is discussed based on the response entropy in Sect. 4.4.

Experiment Strategy and Setup

To realistically determine the behavior of the different PUFs, in particular regarding their reproducibility, they need to be tested under varying conditions. In particular, we consider variations in test chip's supply voltage, $V_{dd} = 1.02 \, \text{V} \ldots 1.32 \, \text{V}$, and environment temperature, $T_{env} = -45\ °C \ldots 85\ °C$, during evaluation. These conditions are created by powering the test chip with a variable power supply and placing it in a climate chamber with temperature control. To comprehensibly assess the PUF's behavior over these intervals, we test it at the four extreme corner conditions:

- The low-low corner (LL) or $\alpha_{LL} = (T_{env} = -45\ °C, V_{dd} = 1.02\ V)$.
- The low-high corner (LH) or $\alpha_{LH} = (T_{env} = -45\ °C, V_{dd} = 1.32\ V)$.
- The high-low corner (HL) or $\alpha_{HL} = (T_{env} = 85\ °C, V_{dd} = 1.02\ V)$.
- The high-high corner (HH) or $\alpha_{HH} = (T_{env} = 85\ °C, V_{dd} = 1.32\ V)$.

In addition, the test chip is also evaluated at the nominal reference condition: $\alpha_{ref} = (T_{env} = 25\ °C, V_{dd} = 1.20\ V)$.

At all considered conditions, all PUF constructions on all $N_{puf} = 192$ test chips are evaluated for all their possible challenges, except for the arbiter PUFs, which are only evaluated on a set of 256 randomly generated challenges. In fact, for the memory-based PUFs, the notion of 'number of challenges' is rather arbitrary since it depends on how many bits one considers to be in a response, which we denote by ℓ_{resp}. In that respect, it is much more natural to detail the total number of generated response bits instead, denoted as $N_{bits} \equiv N_{chal} \times \ell_{resp}$. Since all considered evaluation methods for the delay-based PUFs also generate bitwise responses, we also describe them in this manner. For the arbiter PUFs, we consider all bits generated by all arbiter PUFs at the same time; hence for the basic arbiter PUF experiment, $N_{bits} = 256 \times 256 = 65536$, and for the 2-XOR arbiter PUF, $N_{bits} = 256 \times 128 = 32768$. All response bits are evaluated $N_{meas} = 20$ times under each condition for all PUFs on all chips.

4.3.3 Experimental PUF Uniqueness Results

Uniqueness Quantifiers

As expressed in Definition 8, the uniqueness of a PUF class is determined by the distribution of its inter-distance, in particular at nominal conditions. To study this distribution, we calculate the inter-distances $\mathbf{D}^{inter}_{Exp(\mathcal{P})}$ on the measured responses for all PUF constructions and report the most important statistics on the observed inter-distances in Table 4.2. Since all responses are bitwise, all inter-distances are measured using Hamming distances. However, to make it easier to compare the different PUF constructions, we report them as fractional Hamming distances, i.e. the Hamming distances are expressed as a ratio of the total number of evaluated response bits N_{bits}.

The basic inter-distance statistics we report in Table 4.2 are:

- The sample mean $\mu^{inter}_{\mathcal{P}}$ and the sample standard deviation $\sigma^{inter}_{\mathcal{P}}$, respectively expressing the location and dispersion of the inter-distance distribution.

Table 4.2 Experimental uniqueness results: inter-distance statistics (at nominal condition)

PUF	No.	N_{bits}	μ_P^{inter}	σ_P^{inter}	$P[1\%]_P^{inter}$	$\min_P^{inter}$	$\hat{p}_P^{inter}$
SRAM PUF	0	65536	49.59 %	0.33 %	48.77 %	47.82 %	48.75 %
	1	65536	49.61 %	0.33 %	48.76 %	47.93 %	48.86 %
	2	65536	49.68 %	0.31 %	48.88 %	47.80 %	48.72 %
	3	65536	49.72 %	0.30 %	48.94 %	48.12 %	49.04 %
Latch PUF	0	8192	34.84 %	1.20 %	31.82 %	28.37 %	30.77 %
	1	8192	37.01 %	1.23 %	33.86 %	31.24 %	33.70 %
	2	8192	33.17 %	1.62 %	29.31 %	25.27 %	27.59 %
	3	8192	16.37 %	2.02 %	12.10 %	10.43 %	12.10 %
D Flip-Flop PUF	0	8192	42.35 %	0.83 %	40.36 %	38.14 %	40.70 %
	1	8192	42.42 %	1.01 %	40.15 %	38.20 %	40.76 %
	2	8192	41.88 %	0.89 %	39.80 %	37.87 %	40.43 %
	3	8192	41.20 %	0.87 %	39.15 %	36.95 %	39.50 %
Buskeeper PUF	0	8192	48.88 %	0.71 %	47.12 %	45.65 %	48.27 %
	1	8192	48.92 %	0.69 %	47.23 %	45.67 %	48.28 %
Arbiter PUF (basic)	0	65536	47.13 %	0.44 %	46.13 %	45.51 %	46.43 %
Arbiter PUF (2-XOR)	0	32768	49.74 %	0.29 %	49.07 %	48.40 %	49.71 %
Ring Oscillator PUF (P.C.)	0	2048	49.60 %	1.11 %	47.02 %	44.58 %	49.54 %
Ring Oscillator PUF (L.G.)	0	12544	46.86 %	0.48 %	45.77 %	44.34 %	46.45 %

- The first percentile $P[1\%]_{\mathcal{P}}^{\text{inter}}$ of the samples and the sample minimum $\min_{\mathcal{P}}^{\text{inter}}$. These two order statistics give a good idea of the left tail of the distribution, which is of interest when quantifying identifiability later.

In addition to these standard statistics, we introduce a custom statistic which will be of use later on: the inter-distance binomial probability estimator $\hat{p}_{\mathcal{P}}^{\text{inter}}$. For many applications, we need to make assumptions about the distribution of the inter-distance of a PUF construction. In particular, we often need to extrapolate the observed empirical distribution to very small or large probabilities for which we have no reliable measurements. For efficiency reasons it is important that this be done as accurately as possible. However, any overestimation of the uniqueness could be disastrous for the security requirements of an application and should be avoided at all cost. For extrapolations beyond the observed inter-distance values, we make the assumption that the inter-distance is *binomially distributed* with parameter $\hat{p}_{\mathcal{P}}^{\text{inter}}$. This parameter is chosen to be as large as possible, but sufficiently small such that all three following constraints are met, with $F_{\text{bino}}(x; n, p)$ the cumulative binomial distribution function with parameters n and p, evaluated in x:

1. $F_{\text{bino}}(\mu_{\mathcal{P}}^{\text{inter}} \cdot N_{\text{bits}}; N_{\text{bits}}, \hat{p}_{\mathcal{P}}^{\text{inter}}) \geq 50\%$, i.e. the estimated binomial distribution should produce values smaller than or equal to the observed sample mean with a probability of at least 50 %.
2. $F_{\text{bino}}(P[1\%]_{\mathcal{P}}^{\text{inter}} \cdot N_{\text{bits}}; N_{\text{bits}}, \hat{p}_{\mathcal{P}}^{\text{inter}}) \geq 1\%$, i.e. the estimated binomial distribution should produce values smaller than or equal to the observed first percentile with a probability of at least 1 %. If this is not the case, values smaller than or equal to the first percentile of the samples are unlikely to occur as often as they do in the experiment, which means the estimated binomial distribution is an overestimation.
3. $F_{\text{bino}}(\min_{\mathcal{P}}^{\text{inter}} \cdot N_{\text{bits}}; N_{\text{bits}}, \hat{p}_{\mathcal{P}}^{\text{inter}}) \geq 10^{-6}$, i.e. the estimated binomial distribution should produce values smaller than or equal to the observed sample minimum with a probability of at least 10^{-6} (the total number of observed samples is in the order of one million). If this is not the case, the observed sample minimum is unlikely to occur in the experiment according to the estimated binomial distribution, which is hence an overestimation.

The largest value for $\hat{p}_{\mathcal{P}}^{\text{inter}}$ meeting all these three constraints is computed for all the PUF constructions and reported in Table 4.2. When the inter-distance is approximately binomially distributed, the value for $\hat{p}_{\mathcal{P}}^{\text{inter}}$ should closely match that for $\mu_{\mathcal{P}}^{\text{inter}}$. When this is not the case, the three constraints make sure that the binomial estimation based on $\hat{p}_{\mathcal{P}}^{\text{inter}}$ at least accurately models the left tail of the actual inter-distance distribution, to avoid overestimation for extrapolations to small probabilities.

Discussion on Uniqueness Results

When presented as fractional Hamming distance, the optimal inter-distance is 50 %, which indicates two response bit vectors are maximally uncorrelated. A reported inter-distance sample mean (and binomial probability estimator) close to 50 % hence indicates high uniqueness. In this respect, the SRAM PUF and the buskeeper PUF perform particularly well, whereas the D flip-flip PUF and the latch PUF show slightly reduced uniqueness. In particular latch PUF instance number 3, with an average inter-distance of merely 16 %, shows strikingly little uniqueness, even in comparison to the other latch PUF instances. This is a sign of a possible implementation error in this latch instance. The basic arbiter PUF shows high uniqueness, which is an indication that the full-custom design approach succeeded in minimizing the arbiter PUF bias. For the 2-XOR arbiter PUF the uniqueness is evidently even higher. Both ring oscillator PUF evaluation methods also offer high uniqueness. The pairwise comparison approach has slightly better uniqueness than the Lehmer-Gray encoding, but the latter method of course produces significantly more response bits from the same amount of oscillators. For all the PUFs, the comparison between the inter-distance sample mean and binomial probability estimator is also an indication of how closely their inter-distance distribution resembles a binomial distribution.

4.3.4 Experimental PUF Reproducibility Results

Reproducibility Quantifiers

The reproducibility of a PUF class is determined by the distribution of its intra-distance (cf. Definition 7), in particular at the most extreme reference corner conditions. To study these distributions, we calculate the intra-distances $\mathbf{D}^{intra}_{Exp(\mathcal{P})}$ on the measured responses for all PUF constructions at the reference condition α_{ref}, and at all four corner conditions, α_{LL}, α_{LH}, α_{HL}, α_{HH}, and report the most important statistics on the observed intra-distances. Tables 4.3, 4.4, 4.5, 4.6 and 4.7 respectively report the intra-distance statistics for the experiments under conditions α_{ref}, α_{LL}, α_{LH}, α_{HL} and α_{HH}.

The basic intra-distance statistics we report in these tables are:

- The sample mean $\mu^{intra}_{\mathcal{P}}$ and the sample standard deviation $\mu^{intra}_{\mathcal{P}}$, respectively expressing the location and dispersion of the intra-distance distributions.
- The 99th percentile $P[99\ \%]^{intra}_{\mathcal{P}}$ of the samples and the sample maximum $\max^{intra}_{\mathcal{P}}$. These two order statistics give a good idea of the right tail of the distribution, which is of interest when quantifying identifiability later.

In addition to these standard statistics, we again introduce a custom statistic for realistically approximating the observed distribution by a binomial distribution: the intra-distance binomial probability estimator $\hat{p}^{intra}_{\mathcal{P}}$. This time, this estimator is constrained to at least accurately model the right tail of the distribution to avoid underestimation of the intra-distance distribution at high values. The reported values for $\hat{p}^{intra}_{\mathcal{P}}$ are computed to be as small as possible, but sufficiently large such that all three following constraints are met.

Table 4.3 Intra-distance statistics at nominal condition: $\alpha_{ref} = (T_{env} = 25\ °C,\ V_{dd} = 1.20\ V)$

PUF	No.	N_{bits}	μ_P^{intra}	σ_P^{intra}	$P[99\%]_P^{intra}$	$\max_P^{intra}$	$\hat{p}_P^{intra}$
SRAM PUF	0	65536	5.46 %	0.14 %	5.77 %	6.06 %	5.63 %
	1	65536	5.46 %	0.14 %	5.76 %	6.02 %	5.59 %
	2	65536	5.46 %	0.14 %	5.76 %	5.96 %	5.55 %
	3	65536	5.47 %	0.14 %	5.76 %	6.00 %	5.58 %
Latch PUF	0	8192	2.61 %	0.24 %	3.16 %	3.65 %	2.77 %
	1	8192	2.78 %	0.25 %	3.37 %	3.92 %	3.00 %
	2	8192	3.40 %	0.34 %	4.28 %	5.16 %	4.10 %
	3	8192	2.64 %	0.55 %	4.13 %	5.27 %	4.20 %
D Flip-Flop PUF	0	8192	3.54 %	0.23 %	4.08 %	4.61 %	3.61 %
	1	8192	3.76 %	0.53 %	5.71 %	14.47 %	12.70 %
	2	8192	3.50 %	0.24 %	4.10 %	4.88 %	3.85 %
	3	8192	3.45 %	0.23 %	4.00 %	4.74 %	3.72 %
Buskeeper PUF	0	8192	4.16 %	0.24 %	4.72 %	5.21 %	4.22 %
	1	8192	4.17 %	0.24 %	4.74 %	5.21 %	4.23 %
Arbiter PUF (basic)	0	65536	3.04 %	0.08 %	3.23 %	3.39 %	3.07 %
Arbiter PUF (2-XOR)	0	32768	5.89 %	0.15 %	6.26 %	6.57 %	5.95 %
Ring Oscillator PUF (P.C.)	0	2048	1.53 %	0.39 %	2.44 %	3.13 %	1.80 %
Ring Oscillator PUF (L.G.)	0	12544	3.56 %	0.63 %	4.68 %	5.26 %	4.38 %

Table 4.4 Intra-distance statistics at LL corner conditions: $\alpha_{LL} = (T_{env} = -40\,^\circ C,\ V_{dd} = 1.02\,V)$

PUF	No.	N_{bits}	$\mu_{\mathcal{P};\alpha_{LL}}^{intra}$	$\sigma_{\mathcal{P};\alpha_{LL}}^{intra}$	$P[99\%]_{\mathcal{P};\alpha_{LL}}^{intra}$	$\max_{\mathcal{P};\alpha_{LL}}^{intra}$	$\hat{p}_{\mathcal{P};\alpha_{LL}}^{intra}$
SRAM PUF	0	65536	7.36 %	0.19 %	7.79 %	8.02 %	7.55 %
	1	65536	7.33 %	0.19 %	7.81 %	8.11 %	7.62 %
	2	65536	7.33 %	0.19 %	7.80 %	8.09 %	7.60 %
	3	65536	7.34 %	0.19 %	7.83 %	8.11 %	7.62 %
Latch PUF	0	8192	23.10 %	1.92 %	26.58 %	28.21 %	25.91 %
	1	8192	23.36 %	1.73 %	26.89 %	28.35 %	26.04 %
	2	8192	15.85 %	1.11 %	18.73 %	19.58 %	17.76 %
	3	8192	9.34 %	1.56 %	14.62 %	17.11 %	15.22 %
D Flip-Flop PUF	0	8192	12.79 %	0.91 %	15.26 %	16.61 %	14.74 %
	1	8192	15.61 %	4.06 %	29.60 %	32.83 %	30.41 %
	2	8192	12.56 %	1.10 %	15.53 %	16.70 %	14.82 %
	3	8192	12.35 %	1.20 %	16.48 %	17.86 %	15.93 %
Buskeeper PUF	0	8192	9.68 %	0.41 %	10.62 %	11.39 %	9.86 %
	1	8192	9.89 %	0.39 %	10.80 %	11.44 %	10.04 %
Arbiter PUF (basic)	0	65536	7.41 %	0.23 %	7.98 %	8.25 %	7.75 %
Arbiter PUF (2-XOR)	0	32768	13.72 %	0.40 %	14.70 %	15.14 %	14.25 %
Ring Oscillator PUF (P.C.)	0	2048	3.75 %	0.47 %	4.83 %	5.62 %	3.88 %
Ring Oscillator PUF (L.G.)	0	12544	9.01 %	0.47 %	10.35 %	11.17 %	9.89 %

Table 4.5 Intra-distance statistics at LH corner conditions: $\alpha_{LH} = (T_{env} = -40\,°C, V_{dd} = 1.32\,V)$

PUF	No.	N_{bits}	$\mu^{intra}_{\mathcal{P};\alpha_{LH}}$	$\sigma^{intra}_{\mathcal{P};\alpha_{LH}}$	$P[99\%]^{intra}_{\mathcal{P};\alpha_{LH}}$	$\max^{intra}_{\mathcal{P};\alpha_{LH}}$	$\hat{p}^{intra}_{\mathcal{P};\alpha_{LH}}$
SRAM PUF	0	65536	7.46 %	0.20 %	7.91 %	8.15 %	7.67 %
	1	65536	7.44 %	0.20 %	7.94 %	8.28 %	7.78 %
	2	65536	7.44 %	0.20 %	7.93 %	8.28 %	7.78 %
	3	65536	7.44 %	0.20 %	7.96 %	8.25 %	7.75 %
Latch PUF	0	8192	23.38 %	1.92 %	26.92 %	28.13 %	25.82 %
	1	8192	23.66 %	1.74 %	27.27 %	28.65 %	26.33 %
	2	8192	15.51 %	0.99 %	17.96 %	19.20 %	17.21 %
	3	8192	13.70 %	2.34 %	19.81 %	24.04 %	21.86 %
D Flip-Flop PUF	0	8192	12.90 %	0.91 %	15.33 %	16.92 %	15.03 %
	1	8192	15.68 %	4.08 %	29.63 %	33.02 %	30.60 %
	2	8192	12.64 %	1.09 %	15.59 %	16.75 %	14.87 %
	3	8192	12.42 %	1.20 %	16.52 %	17.76 %	15.83 %
Buskeeper PUF	0	8192	9.77 %	0.41 %	10.72 %	11.51 %	9.96 %
	1	8192	10.02 %	0.39 %	10.93 %	11.65 %	10.16 %
Arbiter PUF (basic)	0	65536	5.41 %	0.22 %	6.20 %	6.94 %	6.48 %
Arbiter PUF (2-XOR)	0	32768	10.23 %	0.39 %	11.58 %	12.88 %	12.02 %
Ring Oscillator PUF (P.C.)	0	2048	3.03 %	0.42 %	4.05 %	4.98 %	3.19 %
Ring Oscillator PUF (L.G.)	0	12544	7.81 %	0.41 %	8.69 %	9.32 %	8.15 %

Table 4.6 Intra-distance statistics at HL corner conditions: $\alpha_{HL} = (T_{env} = 85\,°C,\ V_{dd} = 1.02\,V)$

PUF	No.	N_{bits}	$\mu_{\mathcal{P};\alpha_{HL}}^{\text{intra}}$	$\sigma_{\mathcal{P};\alpha_{HL}}^{\text{intra}}$	$P[99\,\%]_{\mathcal{P};\alpha_{HL}}^{\text{intra}}$	$\max_{\mathcal{P};\alpha_{HL}}^{\text{intra}}$	$\hat{p}_{\mathcal{P};\alpha_{HL}}^{\text{intra}}$
SRAM PUF	0	65536	7.28 %	0.15 %	7.63 %	7.95 %	7.46 %
	1	65536	7.33 %	0.15 %	7.70 %	8.01 %	7.52 %
	2	65536	7.33 %	0.16 %	7.72 %	7.99 %	7.50 %
	3	65536	7.34 %	0.14 %	7.68 %	7.99 %	7.50 %
Latch PUF	0	8192	10.62 %	1.10 %	14.47 %	17.96 %	16.02 %
	1	8192	11.99 %	1.24 %	15.97 %	19.43 %	17.43 %
	2	8192	12.51 %	1.20 %	15.66 %	17.04 %	15.15 %
	3	8192	7.90 %	1.45 %	12.42 %	13.94 %	12.21 %
D Flip-Flop PUF	0	8192	18.10 %	0.79 %	20.11 %	21.03 %	19.11 %
	1	8192	17.95 %	1.66 %	21.89 %	23.71 %	21.54 %
	2	8192	18.27 %	0.78 %	20.23 %	21.08 %	19.23 %
	3	8192	18.15 %	0.81 %	19.96 %	20.84 %	18.96 %
Buskeeper PUF	0	8192	17.71 %	0.90 %	20.06 %	21.14 %	19.07 %
	1	8192	17.60 %	0.91 %	19.95 %	21.03 %	18.97 %
Arbiter PUF (basic)	0	65536	5.23 %	0.21 %	5.91 %	6.29 %	5.86 %
Arbiter PUF (2-XOR)	0	32768	9.90 %	0.37 %	11.13 %	11.83 %	11.00 %
Ring Oscillator PUF (P.C.)	0	2048	2.84 %	0.40 %	3.81 %	4.54 %	2.98 %
Ring Oscillator PUF (L.G.)	0	12544	7.11 %	0.42 %	8.10 %	9.19 %	8.03 %

Table 4.7 Intra-distance statistics at HH corner conditions: $\alpha_{HH} = (T_{env} = 85\,°C,\ V_{dd} = 1.32\,V)$

PUF	No.	N_{bits}	$\mu^{intra}_{\mathcal{P};\alpha_{HH}}$	$\sigma^{intra}_{\mathcal{P};\alpha_{HH}}$	$P[99\,\%]^{intra}_{\mathcal{P};\alpha_{HH}}$	$\max^{intra}_{\mathcal{P};\alpha_{HH}}$	$\hat{p}^{intra}_{\mathcal{P};\alpha_{HH}}$
SRAM PUF	0	65536	7.28 %	0.15 %	7.63 %	7.87 %	7.39 %
	1	65536	7.33 %	0.15 %	7.69 %	8.00 %	7.51 %
	2	65536	7.32 %	0.15 %	7.70 %	7.95 %	7.47 %
	3	65536	7.33 %	0.14 %	7.66 %	8.08 %	7.59 %
Latch PUF	0	8192	10.60 %	1.08 %	14.42 %	17.81 %	15.88 %
	1	8192	11.98 %	1.21 %	15.86 %	19.07 %	17.08 %
	2	8192	12.71 %	1.27 %	16.08 %	19.10 %	17.12 %
	3	8192	7.37 %	1.28 %	10.82 %	12.04 %	10.42 %
D Flip-Flop PUF	0	8192	17.89 %	0.79 %	19.90 %	20.87 %	18.90 %
	1	8192	17.77 %	1.67 %	21.79 %	23.38 %	21.22 %
	2	8192	18.12 %	0.78 %	20.06 %	20.84 %	19.06 %
	3	8192	17.98 %	0.81 %	19.81 %	20.48 %	18.82 %
Buskeeper PUF	0	8192	17.48 %	0.89 %	19.81 %	20.70 %	18.82 %
	1	8192	17.38 %	0.89 %	19.69 %	20.79 %	18.73 %
Arbiter PUF (basic)	0	65536	5.34 %	0.24 %	5.92 %	6.46 %	6.02 %
Arbiter PUF (2-XOR)	0	32768	10.11 %	0.44 %	11.14 %	12.13 %	11.30 %
Ring Oscillator PUF (P.C.)	0	2048	3.27 %	0.42 %	4.30 %	5.22 %	3.41 %
Ring Oscillator PUF (L.G.)	0	12544	8.35 %	0.53 %	9.89 %	10.79 %	9.54 %

1. $F_{\text{bino}}(\mu_{\mathcal{P}}^{\text{intra}} \cdot N_{\text{bits}}; N_{\text{bits}}, \hat{p}_{\mathcal{P}}^{\text{intra}}) \leq 50\%$, i.e. the estimated binomial distribution should produce values larger than or equal to the observed sample mean with a probability of at least 50 %.

2. $F_{\text{bino}}(P[99\%]_{\mathcal{P}}^{\text{intra}} \cdot N_{\text{bits}}; N_{\text{bits}}, \hat{p}_{\mathcal{P}}^{\text{intra}}) \leq 99\%$, i.e. the estimated binomial distribution should produce values larger than or equal to the observed 99th percentile with a probability of at least 1 %. If this is not the case, values larger than or equal to the 99th percentile of the samples are unlikely to occur as often as they do in the experiment, which means the estimated binomial distribution is an underestimation.

3. $F_{\text{bino}}(\max_{\mathcal{P}}^{\text{intra}} \cdot N_{\text{bits}}; N_{\text{bits}}, \hat{p}_{\mathcal{P}}^{\text{intra}}) \leq 1 - 10^{-6}$, i.e. the estimated binomial distribution should produce values larger than or equal to the observed sample maximum with a probability of at least 10^{-6} (the number of samples in the experiment is in the order of one million). If this is not the case, the observed sample maximum is unlikely to occur in the experiment according to the estimated binomial distribution, which is hence an underestimation.

When the intra-distance is approximately binomially distributed, the value for $\hat{p}_{\mathcal{P}}^{\text{intra}}$ should closely match that for $\mu_{\mathcal{P}}^{\text{intra}}$. When this is not the case, the three constraints make sure that the binomial estimation based on $\hat{p}_{\mathcal{P}}^{\text{intra}}$ at least accurately models the right tail of the actual intra-distance distribution, to avoid underestimation for extrapolations to larger probabilities.

Discussion on Reproducibility Results

The intra-distance statistics reported in Tables 4.3 to 4.7 clearly show that the evaluation conditions impact the reproducibility of a PUF. For integration in a realistic application, we are particularly interested in the worst-case reproducibility behavior over all considered conditions. In Table 4.8, we summarize the worst-case results from Tables 4.3 to 4.7, i.e. the statistics from these tables which show the largest intra-distances. Note that different worst-case results can come from different conditions.

Studying the worst-case intra-distance statistics in Table 4.8, we see that of all the memory-based PUF constructions the SRAM PUF is particularly reproducible, with even worst-case maximal intra-distances smaller than 10 %. The buskeeper, D flip-flop and latch PUFs have a considerably worse reproducibility. We also remark two outlying behaviors:

1. Latch PUF instances numbers 2 and 3 show lower intra-distances than the other two, which is a side effect of them already having low uniqueness. This is most likely a result of an implementation fault in these instances, which we also already spotted based on the outlying uniqueness results. Note that latch instances numbers 2 and 3 deploy the rather complex scan chain-based read-out technique, as opposed to the other two which use a multiplexer tree, which is likely the cause of this problem.

Table 4.8 Worst-case intra-distance statistics over all four corners (α_{wc})

PUF	No.	N_{bits}	$\mu^{intra}_{\mathcal{P};\alpha_{wc}}$	$P[99\%]^{intra}_{\mathcal{P};\alpha_{wc}}$	$max^{intra}_{\mathcal{P};\alpha_{wc}}$	$\hat{p}^{intra}_{\mathcal{P};\alpha_{wc}}$
SRAM PUF	0	65536	7.46 %	7.91 %	8.15 %	7.67 %
	1	65536	7.44 %	7.94 %	8.28 %	7.78 %
	2	65536	7.44 %	7.93 %	8.28 %	7.78 %
	3	65536	7.44 %	7.96 %	8.25 %	7.75 %
Latch PUF	0	8192	23.38 %	26.92 %	28.21 %	25.91 %
	1	8192	23.66 %	27.27 %	28.65 %	26.33 %
	2	8192	15.85 %	18.73 %	19.58 %	17.76 %
	3	8192	13.70 %	19.81 %	24.04 %	21.86 %
D Flip-Flop PUF	0	8192	18.10 %	20.11 %	21.03 %	19.11 %
	1	8192	17.95 %	29.63 %	33.02 %	30.60 %
	2	8192	18.27 %	20.23 %	21.08 %	19.23 %
	3	8192	18.15 %	19.96 %	20.84 %	18.96 %
Buskeeper PUF	0	8192	17.71 %	20.06 %	21.14 %	19.07 %
	1	8192	17.60 %	19.95 %	21.03 %	18.97 %
Arbiter PUF (basic)	0	65536	7.41 %	7.98 %	8.25 %	7.75 %
Arbiter PUF (2-XOR)	0	32768	13.72 %	14.70 %	15.14 %	14.25 %
Ring Oscillator PUF (P.C.)	0	2048	3.75 %	4.83 %	5.62 %	3.88 %
Ring Oscillator PUF (L.G.)	0	12544	9.01 %	10.35 %	11.17 %	9.89 %

2. D flip-flop PUF instance number one shows considerably larger intra-distances than the other three instances. The cause for this is unknown, but is likely also due to an implementation fault.

To prevent these outlying results from affecting this objective comparison between different PUF constructions, we will ignore them in all following analysises.

The delay-based PUFs also show relatively good reproducibility. The worst-case statistics of the basic arbiter PUF are nearly identical to those of the SRAM PUF. For the 2-XOR arbiter PUF, the intra-distances get approximately twice as large, which follows from their construction: when one of the two XOR-ed arbiter PUFs produces a faulty bit, the XOR-ed result will also be wrong. The ring oscillator PUF response bits based on pairwise comparison of frequencies are extremely reproducible, which proves the strength of this method. The Lehmer-Gray encoding method has worse reproducibility than the pairwise comparison method, but is still fairly good with a worst-case average smaller than 10 %.

4.4 Assessing Entropy

For many applications, and in particular for PUF-based key generation, it is important to accurately estimate the *entropy* of a random PUF response. Entropy is a function of the distribution of a random variable and expresses the amount of uncertainty one has about the outcome of the random variable. In the case of PUF responses, it represents a generalized and unconditional upper bound on the average predictability of an unobserved random outcome Y of a response evaluation. However, in general it is also very difficult or even impossible to calculate the entropy of a PUF response exactly. In the end, the distribution of most PUF responses is determined by very complex and even chaotic physical processes, and it cannot be learned in the complete detail which is required to calculate its entropy exactly. Typically, only estimated upper bounds on the underlying entropy can be provided. These bounds are either derived from a high-level physical model of the PUF construction, or based on the experimental data one observes.

In this section, we will present increasingly tighter upper bounds on the entropy of a PUF response based on increasingly more powerful adversary models, i.e. adversaries which gain more and more insight into the underlying distribution of the PUF's responses. We do this for all eight PUF constructions studied on the test chip and we compute entropy bounds based on the experimentally observed response distributions of these PUFs.

4.4.1 Adversary Models and Basic Entropy Bounds

In the following, Y^n represents a random bit vector of length n, and $Y_i \leftarrow \{0, 1\}$ is a binary random variable whose distribution is completely determined by

$p_i \stackrel{\Delta}{=} \Pr(Y_i = 1)$. We also use the following notation for the conditional distribution of Y_i conditioned on the previous bits $Y^{(i-1)} = (Y_1, \ldots, Y_{(i-1)})$: $p_{i|y^{(i-1)}} \stackrel{\Delta}{=}$ $\Pr(Y_i = 1|Y^{(i-1)} = y^{(i-1)})$. An overview of the general notions of probability theory and information theory used in this section is found in Appendix A.

Completely Ignorant Adversary

An adversary which is completely ignorant of the underlying distribution of the responses can make no better prediction than just guessing every bit completely at random. To him, it looks as if the PUF response has full entropy. Based on this adversary, we can introduce the following trivial response entropy bound:

$$H(Y^n) \leq n.$$

Using entropy density, this bound is denoted as

$$\rho_{\text{ignorant}}(Y^n) \stackrel{\Delta}{=} 100\,\%,$$

such that $\rho(Y^n) \leq \rho_{\text{ignorant}}(Y^n)$. This *ignorant entropy bound* is very trivial, but we include it nonetheless for completeness and to detail the manner in which we will discuss the following bounds.

Adversary Knows Global Bias

The most basic deviation from a completely uniform and independent distribution of the response bits is caused by an overall global bias, i.e. on average every bit is more likely to be either '0' or '1'. Such a global bias in the response Y^n can be expressed as:

$$p_{\text{globalbias}} = \frac{1}{n}\mathsf{E}\left[\sum_{i=1}^{n} Y_i\right].$$

An adversary with knowledge of this global bias can make better than random predictions by guessing in favor of the bias, i.e. if $p_{\text{globalbias}} < 50\,\%$ it predicts a '0' and else a '1'. To the adversary, it looks as if all PUF response bits are independent and identically distributed (i.i.d.) according to a Bernoulli distribution with parameter $p_{\text{globalbias}}$. Based on such an adversary, the following response entropy bound is introduced:

$$H(Y^n) \leq n \cdot h(p_{\text{globalbias}}).$$

This *global bias entropy bound* is expressed using entropy density as $\rho(Y^n) \leq \rho_{\text{globalbias}}(Y^n)$, with:

$$\rho_{\text{globalbias}}(Y^n) \stackrel{\Delta}{=} h(p_{\text{globalbias}}).$$

Adversary Knows Bit-Dependent Bias

In a more realistic setting, every bit position in a PUF response vector will have its own bias, as expressed by $p_i = \Pr(y_i = 1)$. An adversary knowing these individual bit-dependent biases can make a more accurate prediction by guessing individual bits in favor of these biases. To the adversary, it looks as if all PUF response bits are independently, but no longer identically distributed, with each bit sampled from its own Bernoulli distribution with parameter p_i. Taking into account this adversary, the response entropy bound can be refined further:

$$H(Y^n) \le \sum_{i=1}^{n} h(p_i).$$

Again we rewrite this *bit-dependent bias entropy bound* using entropy density as $\rho(Y^n) \le \rho_{\text{bitbias}}(Y^n)$, with:

$$\rho_{\text{bitbias}}(Y^n) \stackrel{\Delta}{=} \frac{1}{n} \sum_{i=1}^{n} h(p_i).$$

Adversary Knows Inter-Bit Dependencies

In the previous three adversary models, we moved from an adversary that sees a PUF response as an i.i.d. uniformly random bit vector to one that observes it as a vector of independently distributed bits which are no longer uniform or identically distributed. The next improvement to the adversary model would be to give it insight into the dependencies between different response bits, i.e. it no longer assumes that the response bits are completely independently distributed. It is clear that full knowledge of all inter-bit dependencies is generally unattainable since that would give the complete and exact distribution of the responses. Instead, we assume an adversary which has a certain realistic yet only partial model of the inter-bit dependencies. We consider two such partial dependency models:

1. The adversary has insight into the pairwise joint distributions $p(y_i, y_j)$ of all possible pairs of response bits in Y^n. This is a natural extension of insight into the bit-dependent bias of individual bits.
2. The adversary has partial insight into the conditional distribution $p(y_i | y^{(i-1)})$ of a response bit Y_i given the observation of the previous response bits $Y^{(i-1)} = y^{(i-1)}$. This is typically the case when the adversary deploys a successful next-bit modeling attack on response bits, with a prediction model which is trained on earlier observed responses.

We discuss both these adversarial models separately.

Adversary Knows Pairwise Joint Distributions

When an adversary knows all pairwise joint distributions between the response bits in Y^n, the response entropy bound is further lowered to:

$$H(Y^n) \leq \sum_{i=1}^{n} h(p_i) - \sum_{i=1}^{n-1} I(Y_i; Y_{i+1}).$$

We call this the *pairwise joint distribution entropy bound* and using entropy density, we write $\rho(Y^n) \leq \rho_{\text{pairjoint}}(Y^n)$, with:

$$\rho_{\text{pairjoint}}(Y^n) \overset{\triangle}{=} \frac{1}{n}\left(\sum_{i=1}^{n} h(p_i) - \sum_{i=1}^{n-1} I(Y_i; Y_{i+1}) \right).$$

The mutual information values, i.e. the information shared by consecutive pairs of random bits, are subtracted from the bit-dependent bias entropy bound since they are in a way counted twice. The mutual information between two consecutive random bits can be computed from their pairwise joint distribution. Note that the subtracted amount is dependent on the way the individual random bit variables are ordered in Y^n, since the mutual information is computed over consecutive pairs (Y_i, Y_{i+1}) from $Y^n = (Y_1, \ldots, Y_n)$. Without loss of generalization, we assume that the bits are ordered in such a way as to maximize the subtracted amount. This yields the tightest lower bound on the response entropy.

Adversary Deploys Next-Bit Modeling Attack

In this model, we assume an adversary can perform a modeling attack on the PUF response bits, which after having been trained with $(i-1)$ previously observed response bits, can predict the i-th bit with an average success probability of $p_{\text{model}(i)}$. This results in a response entropy bound of:

$$H(Y^n) \leq \sum_{i=1}^{n} h(p_{\text{model}(i)}).$$

We call this the *model entropy bound*, and in terms of entropy density this becomes $\rho(Y^n) \leq \rho_{\text{model}}(Y^n)$, with:

$$\rho_{\text{model}}(Y^n) \overset{\triangle}{=} \frac{1}{n} \sum_{i=1}^{n} h(p_{\text{model}(i)}).$$

The tightness of this bound depends on the strength of the assumed model, and hence on the information and computational power available to the adversary in order to build this model. The goal of the model is to exploit dependencies between bits in order to make a better than random prediction for the next bit. In gen-

eral, the model's success rate gets better and better as it is trained on more responses, i.e. $p_{model(i)}$ increases with i. This also means that $\rho_{model}(Y^n)$, unlike
for the previously discussed bounds, is not a constant, but is dependent on n. In
general, $\rho_{model}(Y^n)$ decreases for increasing n. For example, if a model produces
near-perfect predictions after having been trained with a large number of observed
response bits, all following response bits will no longer contribute any meaningful
entropy since they are perfectly predictable. Hence, producing more response bits
will only increase the response length and not the entropy, i.e. the entropy density
of the response decreases.

4.4.2 Entropy-Bound Estimations Based on Experimental Results

Next, we evaluate the different entropy bounds on the measured responses obtained
from the performed experiments which were discussed in Sect. 4.3. All entropy
bounds are expressed using entropy density, allowing us to make an easy comparison
between different PUF constructions.

Bound Estimation Strategy

For all memory-based PUF instances, we consider a response bit vector containing
$n = 5000$ bits to estimate the following entropy bounds:

- The ignorant entropy bound is trivially equal to 100 % for all PUFs.
- For the global bias entropy bound, we estimate the global bias by computing the
  response bit sample mean over all 5000 considered bits on all 192 devices.
- For the bit-dependent bias entropy bound, we estimate the bit-dependent biases
  by computing the response bit sample mean over all 192 devices for every bit
  individually.
- For the pairwise distribution entropy bound, we estimate the pairwise joint distributions of all possible pairs of the considered response bits, by counting the
  occurrences of each of the four possible outcomes $(0, 0)$, $(0, 1)$, $(1, 0)$ or $(1, 1)$
  for each considered pair on all 192 devices.

We do not consider the model entropy bound for memory-based PUFs, as no successful modeling attacks on memory-based PUFs are known. All bound estimations
are done based on responses measured at nominal conditions.

For the delay-based PUF instances, we differentiate between the arbiter and the
ring oscillator PUFs. For both response generation methods of the ring oscillator PUF, we perform the same bound estimations as for the memory-based PUFs,
only on different sizes of response bit vectors, respectively $n = 2048$ for the pairwise comparison method and $n = 12544$ for the Lehmer-Gray method. No model-
building attacks for these ring oscillator PUFs are known.

For the arbiter PUF we take a slightly different approach. The first four bounds are computed for response bit vectors of only $n = 256$ bits, for each of the 256 arbiter PUF instances and 128 2-XOR arbiter PUF instances on the test chip individually, but we report only the results for the worst observed instance. Besides these four bounds, we also consider the model entropy bound for both types of arbiter PUFs. This is discussed in more detail in Sect. 4.4.3.

Bound Estimation Results

The results for the estimations of the ignorant entropy bound, the global bias entropy bound, the bit-dependent bias entropy bound and the pairwise distribution entropy bound are presented in Table 4.9. From these results, it is evident that these four estimates represent consecutively tighter upper bounds on the real response entropy.

4.4.3 Modeling Attacks on Arbiter PUFs

From the initial introduction of arbiter PUFs, it was recognized that they are susceptible to modeling attacks. This is a result of the reduced complexity of the dependency between arbiter PUF challenges and responses. For the basic arbiter PUF it was made clear, e.g. by Lee et al. [75], that this dependency is to a high level of accuracy even a linear system. In that case, the unpredictability of the responses results only from the unknown parameters of this underlying system, since once they are learned every response bit is easily predictable with high accuracy. A modeling attack attempts to estimate the unknown model parameters as a function of observed challenge-response pairs. For more advanced arbiter PUF constructions, e.g. the 2-XOR arbiter PUF, the underlying model becomes more complex, but as shown by Rührmair et al. [118], it is still modelable with more advanced modeling techniques.

Modeling with Machine Learning Techniques

A particularly interesting set of modeling techniques are based on machine learning [99]. Machine learning algorithms are able to automatically learn complex behavior and unknown model parameters by generalizing on presented training examples. Another strong motivation for using machine learning algorithms in modeling attacks is that they are generic, i.e. they have the ability to learn any complex behavior and are not a priori restricted to a particular model description (e.g. a linear model).

In [55], we apply two basic machine learning techniques to our experimental arbiter PUF results to test for modelability: (i) artificial neural networks or ANNs [99],

Table 4.9 Entropy density upper-bound estimations

PUF	No.	n	$\rho_{\mathrm{ignorant}}(Y^n)$	$\rho_{\mathrm{globalbias}}(Y^n)$	$\rho_{\mathrm{bitbias}}(Y^n)$	$\rho_{\mathrm{pairjoint}}(Y^n)$
SRAM PUF	0	5000	100.00 %	99.99 %	99.04 %	94.11 %
	1	5000	100.00 %	99.99 %	99.05 %	94.09 %
	2	5000	100.00 %	99.99 %	99.14 %	94.18 %
	3	5000	100.00 %	99.99 %	99.20 %	94.27 %
Latch PUF	0	5000	100.00 %	80.01 %	77.12 %	71.92 %
	1	5000	100.00 %	81.69 %	79.56 %	74.36 %
	2	5000	100.00 %	91.88 %	90.91 %	85.84 %
	3	5000	100.00 %	59.22 %	51.05 %	45.51 %
D Flip-Flop PUF	0	5000	100.00 %	88.88 %	88.36 %	83.36 %
	1	5000	100.00 %	89.13 %	88.65 %	83.70 %
	2	5000	100.00 %	88.00 %	87.55 %	82.57 %
	3	5000	100.00 %	86.77 %	86.34 %	81.34 %
Buskeeper PUF	0	5000	100.00 %	98.82 %	97.94 %	93.00 %
	1	5000	100.00 %	99.03 %	98.02 %	93.05 %
Arbiter PUF (basic)	0	256	100.00 %	94.27 %	92.79 %	89.62 %
Arbiter PUF (2-XOR)	0	256	100.00 %	99.39 %	98.72 %	95.82 %
Ring Oscillator PUF (P.C.)	0	2048	100.00 %	99.99 %	99.04 %	94.69 %
Ring Oscillator PUF (L.G.)	0	12544	100.00 %	99.92 %	94.87 %	86.63 %

Fig. 4.4 Average success rate of a machine-learning attack on basic and 2-XOR arbiter PUFs. The success rate $p_{model(n)}$ presents the probability of correctly predicting the n-th response bit after having been trained with $(n-1)$ previously observed bits

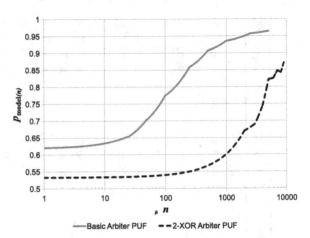

and (ii) support-vector machines or SVMs [26]. Both techniques have been demonstrated to be able to effectively model basic arbiter PUFs, respectively by Gassend et al. [43] and Rührmair et al. [118]. However, we apply these attacks on PUF responses resulting from a modern implementation (65 nm CMOS), as opposed to these earlier results which work with older technologies [43] or only with simulated data [118]. For this analysis of the model entropy bound, we are mainly interested in the results of these modeling attacks. For more details on their implementation, we refer to [55].

Modeling Results

Using both ANN and SVM, we were able to successfully model both basic and 2-XOR arbiter PUFs. Both ANN and SVM first take a number of known arbiter PUF challenge-response pairs which they use to train their model. Afterwards, their modeling performance is evaluated by their success rate of accurately predicting unobserved responses when presented with a challenge from a large test set. It is evident that, the more training examples a machine learning algorithm is allowed to use, the better its modeling accuracy becomes. In Fig. 4.4, we summarize the outcome of our machine learning modeling attacks on the experimental data from the basic and the 2-XOR arbiter PUFs. It shows the average success rate ($p_{model(i)}$) of the best machine learning technique, ANN or SVM, after having been trained with ($i - 1$) earlier observed response bits.

We can draw some conclusions on the machine learning results as represented in Fig. 4.4:

- The 2-XOR arbiter PUF is more difficult to model with our techniques than the basic arbiter PUF, as expressed by the larger number of training examples required to achieve the same modeling success rate. This is a result of the challenge-response relation being more complex for the 2-XOR arbiter PUF.

Fig. 4.5 Estimated entropy
density bounds for the basic
and 2-XOR arbiter PUFs,
based on a modeling
adversary deploying the
machine-learning attack
results presented in Fig. 4.4

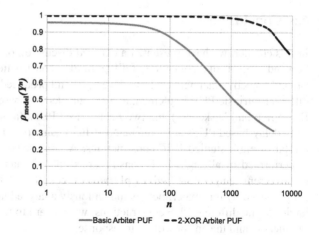

Basic Arbiter PUF   2-XOR Arbiter PUF

- The basic arbiter PUF can be modeled with ≈90 % accuracy after training with ≈500 examples and with ≈95 % accuracy after training with ≈2000 examples. Note that the maximal attainable success rate of any modeling attack is naturally limited by the reproducibility of the considered PUF instance. In that respect, the obtained modeling accuracy of ≈97 % after training with ≈5000 examples can be considered perfect, given that the average intra-distance of the basic arbiter PUF at nominal conditions is about 3 % (cf. Table 4.3). This means that all following response bits do not contribute any entropy except for their reproducibility uncertainty.

- The 2-XOR arbiter PUF can be modeled with ≈75 % accuracy after training with ≈4000 examples and with nearly 90 % after training with 9000 examples.

Modeling Entropy Bound

Using the machine learning modeling attack results as presented in Fig. 4.4, we can calculate the model entropy bound as $\rho_{\text{model}}(Y^n) = \frac{1}{n} \sum_{i=1}^{n} h(p_{\text{model}(i)})$. The resulting entropy bound, as a function of n, is presented in Fig. 4.5.

From Fig. 4.5 we learn that for the basic arbiter PUF, $\rho_{\text{model}}(Y^{100}) < 90$ % and $\rho_{\text{model}}(Y^{1000}) \approx 50$ %. For response bit vectors of more than 1000 bits, the entropy density drops drastically as each additional response bit adds very little entropy. After about 5000 bits, each additional response bit only adds a little noise entropy, which is not useful. For the 2-XOR PUF, the results are less severe with $\rho_{\text{model}}(Y^{5000} < 90$ %) and $\rho_{\text{model}}(Y^{8000} < 80$ %).

As a final remark, we want to make clear that the obtained model entropy bounds are likely not very tight, since our machine-learning modeling attacks are not optimal. More advanced or more fine-tuned modeling techniques are likely to obtain even higher success rates, and hence lower model entropy bounds, than our results, which are based on relatively basic machine learning algorithms.

4.5 Conclusion

In this chapter, we have presented a detailed discussion on the realization, evaluation and analysis of a collection of different PUF constructions in a realistic, practical and objective manner. We have developed and produced a test chip carrying six different intrinsic PUF implementations: four memory-based PUFs (SRAM, latch, D flip-flop and buskeeper) and two delay-based PUFs (arbiter and ring oscillator). Both delay-based PUFs are evaluated (off-line) using two different methods, giving a total of eight studied PUF constructions. A large-scale experimental evaluation is performed of all eight PUF constructions on 192 manufactured test chips under different temperature and supply voltage conditions. The large response data set produced by this experiment is meticulously analyzed in order to assess the behavior of the different PUF constructions, with regard to their reproducibility, their uniqueness and the entropy of their responses.

4.5.1 Summary of PUF Behavior Results

As a final conclusion, we present a concise summary of the most important results related to the PUF behavior of the different intrinsic PUF implementations which we studied in this chapter. We characterize the different PUFs for each of the three analyzed properties (reproducibility, uniqueness and response entropy) in a single quantifier which will be of particular practical use in the following chapters, which discuss PUF-based applications. The three quantifiers we consider are:

1. The intra-distance binomial probability estimator $\hat{p}_{\mathcal{P}}^{\text{intra}}$ as a characterization of reproducibility. This will allow us to accurately and safely estimate the right tail of the intra-distance distribution, i.e. the probability that the intra-distance becomes very large.
2. The inter-distance binomial probability estimator $\hat{p}_{\mathcal{P}}^{\text{inter}}$ as a characterization of uniqueness. This quantifier allows us to accurately and safely estimate the left tail of the inter-distance distribution, i.e. the probability that very low inter-distances occur.
3. The tightest upper bound on the response entropy density $\rho(Y^n)$ as a characterization of response entropy.

For all three quantifiers, we selected the worst-case measured values over all implemented instances of the individual PUF types, respectively from Tables 4.8, 4.2 and 4.9. Note that we discarded all results from latch PUF instances numbers 2 and 3 and from D flip-flop PUF instance number 1, since they exhibit a strong outlier behavior which is likely caused by an implementation error. For the entropy density bound on the basic and 2-XOR arbiter PUFs, we cannot provide a single constant quantifier, since their entropy density bound depends on the considered response length. Instead, we refer to Fig. 4.5, which shows this relation for the machine-learning modeling attacks we performed. Relating to the reported entropy density

Table 4.10 Summary of the most important results on the PUF behavior of the implemented intrinsic PUFs

PUF class $\mathcal{P}$	$\hat{p}_{\mathcal{P}}^{\text{intra}}$	$\hat{p}_{\mathcal{P}}^{\text{inter}}$	$\rho(Y^n) \leq$
SRAM PUF	7.78 %	48.72 %	94.09 %
Latch PUF	26.33 %	30.77 %	71.92 %
D Flip-Flop PUF	19.23 %	39.50 %	81.34 %
Buskeeper PUF	19.07 %	48.27 %	93.00 %
Arbiter PUF (basic)	7.75 %	46.43 %	Fig. 4.5
Arbiter PUF (2-XOR)	14.25 %	49.71 %	Fig. 4.5
Ring Oscillator PUF (P.C.)	3.88 %	49.54 %	94.69 %
Ring Oscillator PUF (L.G.)	9.89 %	46.45 %	86.63 %

results, we also point out that these are merely upper bounds and that the actual response entropy is smaller.

The summary of PUF behavior results in Table 4.10, together with the overview of the area breakdown of the different PUF implementations on the test chip presented in Table 4.1, will be of great value for assessing and optimizing the deployment of these PUF constructions in actual applications.

Chapter 5
PUF-Based Entity Identification and Authentication

5.1 Introduction

5.1.1 Motivation

Due to its combination of uniqueness and reproducibility, a PUF embedded by an entity serves as an identifying feature of that entity, as already intuitively expressed by Definition 9. Moreover, the physical unclonability exhibited by an embedded PUF construction provides even strong security guarantees regarding this expressed identity, which could be used for authentication purposes. However, in order to be of any practical value, the security and robustness of a PUF-based identification or authentication needs to be quantified based on its experimentally verified behavioral characteristics.

Entity Authentication

In information security, the term 'authentication' has a very broad meaning, which often leads to confusion when not described in more detail. First of all, authentication can relate to entities or to data. In the former case one speaks of *entity authentication*, while the latter is called *message authentication*. Since a PUF provides a measure of an entity-specific physical feature, we particularly consider entity authentication in this chapter. Whenever we talk about PUF-based authentication, entity authentication is implied.

Entity authentication by itself is still a catchall for a collection of techniques used to check and be assured of the identity of an entity. The Handbook of Applied Cryptography [96] defines entity authentication as:

Definition 25 An entity authentication technique assures one party, through acquisition of corroborative evidence, of both: (i) the *identity* of a second party involved, and (ii) that the second party was *active* at the time the evidence was created or acquired.

R. Maes, *Physically Unclonable Functions*, DOI 10.1007/978-3-642-41395-7_5,
© Springer-Verlag Berlin Heidelberg 2013

Besides giving convincing proof of its identity, an entity authentication technique also needs to guarantee that the authenticating entity is actively present in the authentication.

Identification

The Handbook of Applied Cryptography [96] treats 'identification' and 'entity authentication' as synonyms. However, in this book, as in many other treatises of the subject, we consider identification to be a related but significantly weaker concept than authentication (see also [96, Remark 10.2]). *Identification* is the mere claiming or stating of its identity, without necessarily presenting any convincing proof thereof. While not strictly a security technique since it doesn't fulfill any meaningful security objective, identification still has very useful qualities:

- Identification is in many cases a necessary precondition for entity authentication and hence an inherent part of most entity authentication techniques. An entity that cannot be identified, cannot be individually authenticated.[1]
- For applications without strict security objectives, identification can be a sufficient condition, e.g. for applications which involve the tracking of products in a closed system.
- In certain situations, identification is sufficient to achieve entity authentication, since the authentication conditions are implicitly met.

For this reason, we will first discuss PUF-based identification in this chapter, before we treat PUF-based authentication.

5.1.2 Chapter Goals

The primary goal of this chapter is to propose practical methods for achieving entity identification and authentication based on the uniqueness and unpredictability of a PUF's challenge-response behavior, and introduce a methodology for quantifying the resulting identification and authentication performance in terms of security and robustness. More specifically, we aim to:

- Study how to use an entity's inherent PUF responses as an identifying feature in an identification system, and how this relates to classic identification based on assigned identities.
- Derive performance metrics for such a PUF-based identification system and apply these on the experimentally derived intrinsic PUF characteristics from Chap. 4, to provide an objective comparison of the identification performance and efficiency of the studied PUFs.
- Develop an entity authentication protocol which: (i) uses a PUF's unique and unpredictable responses directly as an authentication secret, (ii) can be deployed

---

[1]In an anonymous credential scheme, an entity can prove its group membership without revealing its individual identity.

based on existing intrinsic PUFs, and (iii) is sufficiently lightweight to be implemented on resource-constrained devices.

- Derive the performance metrics of the developed authentication scheme and equivalently apply them to the experimental intrinsic PUF characteristics from Chap. 4 to make an objective comparison of their authentication performance.

5.1.3   Chapter Overview

How to safely and reliably identify an entity based on its inherent PUF responses is discussed and analyzed in Sect. 5.2, and a performance overview of the different intrinsic PUFs studied in Chap. 4 is given. In Sect. 5.3, we first describe the operation of an earlier proposed basic PUF-based challenge-response authentication scheme and point out its perceived shortcomings. Based on this analysis, we propose a new and more practical PUF-based mutual authentication scheme and study its authentication performance. Finally, we conclude this chapter in Sect. 5.4.

5.2   PUF-Based Identification

5.2.1   Background: Assigned Versus Inherent Identities

When we compared PUFs to human fingerprints in Sect. 2.1.1, we introduced the concept of an *inherent* identifying feature, i.e. an entity-specific characteristic that arises in the creation process of the entity. As opposed to inherent identities, an entity can also have assigned identities. In the analogy with human beings, this is the distinction we make between fingerprints, which are inherent, and, e.g. a person's name, which is 'assigned' after birth. The inherency of its instance-specific behavior was indicated as one of the key conditions for a construction to be called a PUF.

An inanimate object can also have an assigned identity, e.g. a unique serial number or barcode which is printed on its surface, and in most applications that require entity identification, assigned identities are currently standard practice. In particular for digital silicon chips, unique bit strings which are programmed in a non-volatile memory embedded on the chip were until recently the only way of identifying a specific chip in a digital interaction. With the introduction of silicon PUF technology, it is now possible to also use inherent unique features of a silicon chip for instance identification.

Identity Provisioning Versus Enrollment

Identification techniques based on assigned as well as on inherent identities typically work in two phases. The first phase is different for both types:

- For *assigned identities*, the first phase of any identification technique consists of providing every entity that needs to be identified with a permanent unique identity. We call this the *provisioning phase*.

- For *inherent identities*, the first phase of any identification technique consists of collecting the inherent identities of every entity that needs to be identified. We call this the *enrollment phase*.

The second phase is very similar for both types and consists of an entity presenting its identity, either assigned or inherent, when requested. This is called the *identification phase*.

Practical Advantages of Inherent Identities

The differences between provisioning, for assigned identities, and enrollment, for inherent identities, highlight interesting practical advantages of the latter:

- A unique assigned identity needs to be generated before it is assigned to an entity. To ensure that all generated identities are unique (with high probability), the provisioning party either needs to keep state, e.g. using a monotonic counter, or requires a randomness source, e.g. a true random-number generator. For inherent identities this is not required, since their uniqueness results from the creation process of the entities.
- When assigning an identity to an entity, the provisioning party needs to make permanent (or at least non-volatile) physical changes to each entity. This needs to be supported by the entity's construction. In particular for silicon chips, the inclusion of a non-volatile digital memory can induce a non-negligible additional cost. Evidently, inherent identities do not require additional physical storage capabilities.
- Enrollment (reading an identity) is generally less intrusive than provisioning (writing an identity); hence it can be done faster and with a higher reliability. This is of particular interest for entities created in high-volume manufacturing flows (like silicon chip products), where unit cost is directly affected by the yield and processing time of each manufacturing step.

On the downside, there is no direct control over the actual values taken by inherent identifiers. This is an issue if one wants to assign a meaning to an identifier value, e.g. a serial number which is based on an entity's creation date. For assigned identities, one has absolute control over the identifier values. Another peculiarity of most inherent identities is their so-called *fuzzy nature*, which we discuss next.

5.2.2 Fuzzy Identification

Fuzzy Nature of Inherent Identifiers

A particular trait of most types of inherent identifying features which needs to be dealt with is that they show *fuzzy* random behavior.[2] We say that a random vari-

---

[2]When we use the term 'fuzzy' in this book, we relate to the notion of fuzziness as introduced by Juels and Wattenberg [61] to describe fuzzy commitment, which was later extended to fuzzy

able, like a PUF response or a biometric feature, shows fuzzy behavior if: (i) it is not entirely uniformly distributed, and (ii) it is not perfectly reproducible when measured multiple times. For PUF responses, both fuzzy characteristics are caused by the physical nature of their generation. Random physical processes that introduce entity-specific features during manufacturing are typically not uniformly distributed. Also, as already discussed in detail in Sect. 3.2.2, the response evaluation mechanisms of a PUF construction are subject to physical noise and environmental conditions which cause a non-perfect reproducibility of a PUF response value.

Assigned identities on the other hand are typically not fuzzy. The provisioning party can make sure that the assigned identities are generated from a uniform distribution. Also, once provisioned, an entity can typically reproduce its assigned identity with near-perfect reproducibility.

Fuzzy Identification with a Threshold

The fuzziness of a PUF response is most clearly depicted by its inter- and intra-distance distributions. When we consider binary response vectors and fractional Hamming distance as a distance metric, then perfectly uniformly random responses would have an expected inter-distance of exactly 50 %. It is clear from our literature overview of intrinsic PUF results in Sect. 2.4.8, as well as from our summarized experimental results on intrinsic PUFs in Sect. 4.5, that none of the existing intrinsic PUF constructions meet this condition, although some have an average inter-distance very close to 50 %. Equivalently, no intrinsic PUF exhibits perfect reproducibility with a fixed intra-distance of 0 %, and many are only reproducible up to an average intra-distance of 10 % or even more.

To assess the extent to which a PUF response can be used as an inherent identifier, we need to take its fuzziness into account. This is where the earlier discussed PUF property of *identifiability* comes into play (cf. Definition 9). We defined a PUF class to exhibit identifiability if with high probability its responses' intra-distances are smaller than their inter-distances. In this section, we make this intuitive definition very tangible, by computing the identifying power of a PUF's responses based on their intra- and inter-distance distributions.

Figure 5.1 shows an example of an estimated distribution of the inter- and intra-distances of a PUF's response. For this example, we consider a D flip-flop PUF which produces a 16-bit response. As an estimate for both distributions, we assume a binomial distribution with parameters the binomial probability estimators $\hat{p}_{\mathcal{P}}^{\text{inter}}$ and $\hat{p}_{\mathcal{P}}^{\text{intra}}$ resulting from the experimental analysis from Chap. 4 and summarized in Sect. 4.5. The process by which we computed these estimators guarantees that the assumed binomial distributions provide an accurate estimation, in particular for the

---

vaults by Juels and Sudan [60] and finally to fuzzy extractors by Dodis et al. [32, 33]. We are *not* referring to the homonymous but unrelated use of the word 'fuzzy' as used in fuzzy logic and fuzzy set theory. To avoid any confusion, we will also not use the ambiguous term fuzzy random variable in this context.

Fig. 5.1 Example: estimated inter- and intra-distance distributions for 16-bit responses from the D flip-flop PUF

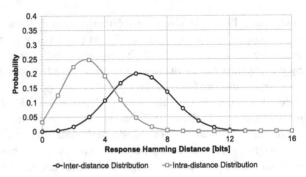

right tail of the intra-distance distribution and for the left tail of the inter-distance distribution. As will become clear, this is specifically the region of interest for most applications.

From Fig. 5.1, it is clear that this PUF construction exhibits some level of identifiability, since the expected intra-distance is noticeably smaller than the expected inter-distance. However, there is also a significant overlap between the curves of the two distributions, which points out the issue of identification based on fuzzy responses. When an observed distance between responses from the enrollment and the identification phase falls in this overlapping region, it can be a result of intra-distance, in which case it is the same entity, or of inter-distance, in which case it concerns a different entity, and there is no way of distinguishing between the two cases. In a practical identification system for fuzzy identities, one needs to determine a rather pragmatic response distance threshold. Distances smaller or equal to this threshold are assumed to be intra-distances between responses from a single entity, while distances above this threshold are assumed to be inter-distances between responses from different entities. We call this threshold the *identification threshold*.

False Acceptance, False Rejection, and Equal Error Rates

During the identification phase of a PUF-based identification system, the generated response of an entity is checked against a list of enrolled responses. When an enrolled response is found whose distance from the presented response is smaller than or equal to the identification threshold, then the entity is identified as the matching entry in the list. It is clear that a fuzzy identification system based on such a pragmatic identification threshold is not 100 % reliable, especially when there is a large overlap between inter- and intra-distance distribution, as for the example in Fig. 5.1. When comparing a presented entity response to a response from the enrollment list, four possible situations can arise:

1. The presented entity is the same entity that produced the enrolled response, and manages to reproduce the enrolled response with an intra-distance smaller than the identification threshold. The presented entity is correctly identified. This is called a *true acceptance*.

2. The presented entity is the same entity that produced the enrolled response, but is not able to reproduce the enrolled response with an intra-distance smaller than the identification threshold. The presented entity is mistakenly rejected. This is called a *false rejection*.
3. The presented entity is not the same entity that produced the enrolled response, but happens (by chance) to produce a response whose inter-distance to the enrolled response is smaller than the identification threshold. The presented entity is mistakenly identified. This is called a *false acceptance*.
4. The presented entity is not the same entity that produced the enrolled response, and it produces a response whose inter-distance is larger than the identification threshold. The presented entity is correctly rejected. This is called a *true rejection*.

It is clear that false rejections and false acceptances are both undesirable for a practical identification system. The probability that a random identification attempt results in one of these cases is respectively expressed as the *false rejection rate* or FRR and as the *false acceptance rate* or FAR of the system. FAR expresses the security of an identification system, since a low FAR means that there is little risk of misidentification which could lead to security issues. FRR on the other hand expresses the robustness or usability of a system, as it expresses the risk of wrongfully rejecting legitimate entities, which would be very impractical. For a usable identification system, both FAR and FRR need to be as small as possible, but it is evident that they cannot be both minimized at the same time. As is often the case, an acceptable trade-off between security and usability needs to be made.

For a given identification system, FAR and FRR depend on the choice for the identification threshold value, which we denote as t_{id}. A high threshold minimizes the risk of a false rejection but increases the likelihood of false acceptances, and vice versa for a low threshold. When the distributions of the inter- and intra-distances of the considered PUF are known, the respective relations between FAR, FRR and t_{id} can be computed:

- FAR is the probability that the inter-distance is smaller than or equal to t_{id}. This is equivalent to the evaluation of the cumulative distribution function of the inter-distance at t_{id}.
- FRR is the probability that the intra-distance is larger than t_{id}. This is equivalent to the complement of the evaluation of the cumulative distribution function of the intra-distance at t_{id}.

For the example presented in Fig. 5.1, we have assumed a binomial distribution for both inter- and intra-distances; hence FAR and FRR become:

$$\text{FAR}(t_{id}) = F_{\text{bino}}\big(t_{id}; 16, \hat{p}_{\mathcal{P}}^{\text{inter}}\big),$$

$$\text{FRR}(t_{id}) = 1 - F_{\text{bino}}\big(t_{id}; 16, \hat{p}_{\mathcal{P}}^{\text{intra}}\big),$$

with $F_{\text{bino}}(t; n, p)$ the cumulative binomial distribution function with parameters n and p evaluated in t, and $\hat{p}_{\mathcal{P}}^{\text{inter}}$ and $\hat{p}_{\mathcal{P}}^{\text{intra}}$ the binomial estimators for the D flip-flop

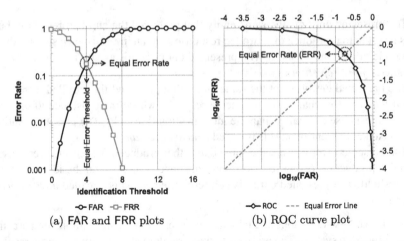

Fig. 5.2 Example: identification metrics for a threshold identification system based on the PUF described by Fig. 5.1

PUF taken from Table 4.10. The resulting FAR and FRR for every identification threshold value between 0 and 16 are plotted in Fig. 5.2a. From this figure it is clear that there is a threshold, and a corresponding error rate, where the plots of FAR and FRR intersect. We call this the *equal error threshold* t_{EER} and the corresponding error rate the *equal error rate* or EER. For discrete distributions, FAR and FRR will never be exactly equal for a discrete threshold, and in that case t_{EER} and EER are defined as:

$$t_{EER} \stackrel{\Delta}{=} \mathrm{argmin}_t\{\max\{\mathsf{FAR}(t), \mathsf{FRR}(t)\}\},$$

and

$$\mathsf{EER} \stackrel{\Delta}{=} \max\{\mathsf{FAR}(t_{EER}), \mathsf{FRR}(t_{EER})\}.$$

The equal error rate is also indicated in Fig. 5.2a.

When designing a PUF-based identification system, the FAR and FRR plots as shown in Fig. 5.2a can be used to find a suitable trade-off meeting the application requirements. This can, but does not need to be the EER, e.g. in some applications more care is given to security than to usability, or vice versa. A more convenient way for assessing the FRR-vs-FAR trade-off is the plot of FRR as a function of FAR as shown in Fig. 5.2b. Such a plot is also called the *receiver-operating characteristic* or ROC plot of the system. In a ROC plot, the EER is found as the intersection with the identity function which we have labelled the *equal error line* in Fig. 5.2b. ROC curves completely summarize the identification performance of an identification system and are particularly useful for comparing the performance of different systems. A more condensed performance qualifier of a particular identification system is given by its EER value.

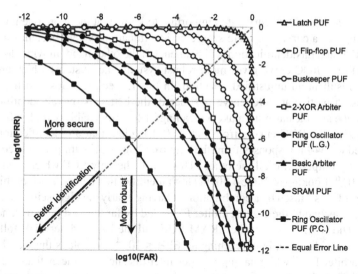

Fig. 5.3 Comparison of ROC curves for identification systems based on 64-bit responses from each of the eight experimentally verified intrinsic PUFs which are summarized by Table 4.10

5.2.3 Identification Performance for Different Intrinsic PUFs

In Sect. 5.2.2 we presented a toy example of a 16-bit D flip-flop PUF identification system and used it to introduce the different performance metrics of fuzzy identification systems. Now, we will use these introduced metrics to objectively compare the identification performance of the different intrinsic PUFs which were experimentally studied in Chap. 4. As in Sect. 5.2.2, we will assume a binomial distribution for both the inter- and intra-distances of these intrinsic PUFs and use the binomial probability estimators as summarized in Table 4.10. This assumption is justified by the fact that these estimators where derived to accurately describe the critical region of both distributions, i.e. the part where both probability mass functions overlap.

We will first compare the identification performance of all considered PUFs for a fixed length response. Afterwards, we compare the required parameters of the different PUFs in order to obtain the same identification performance. Ultimately, we combine these obtained parameters with each PUF's area estimation to generate an objective as possible comparison of identification performance versus silicon area use for each of the intrinsic PUF constructions.

Comparison of ROC Curves for 64-bit Identification with Different PUFs

We consider an identification system based on 64-bit PUF responses and apply the methods introduced in Sect. 5.2.2 to derive the ROC curves for all eight considered intrinsic PUFs, based on their parameters from Table 4.10. All eight ROC curves are plotted on the same graph in Fig. 5.3.

When reading a ROC curve, it is important to know that the more one moves to the left (up) on a curve, the more secure an identification system becomes, i.e. the less likely that a misidentification will happen. On the other hand, moving down (to the right) on a ROC curve gives more robust systems, i.e. systems which are less likely to result in an unjustified rejection. In this respect, the closer to the lower left corner a particular ROC curve is situated, the better its overall identification performance and the easier to make a meaningful security-usability trade-off.

Analyzing the ROC curves from Fig. 5.3, it is clear that an identification system based on 64-bit responses from the ring oscillator PUF with pairwise comparison greatly outperforms all the other PUFs and is the only PUF which obtains an EER $\leq 10^{-6}$. Looking at Table 4.10, the high identification performance of this ring oscillator PUF is caused by a combination of having nearly the highest inter-distance parameter (second to the 2-XOR arbiter PUF) and by far the best intra-distance parameter. The ROC curves of the SRAM PUF and the basic arbiter PUF follow at a considerable distance, both reaching an EER $\leq 10^{-4}$. Notable is the fact that the 2-XOR arbiter PUF, while having a better inter-distance parameter than the basic arbiter PUF, performs significantly worse due to its much worse intra-distance behavior. At the bottom of the ranking we find the latch PUF which performs very weakly. It does not even reach an EER $\leq 10\,\%$, which means that over one in ten identification attempts will either be rejected or misidentified. In fact, looking at the inter- and intra-distance parameters of the latch PUF in Table 4.10, we may conclude that the latch PUF hardly exhibits identifiability (and can hence only barely be called a PUF), since its average intra-distance is only slightly smaller than its average inter-distance.

Comparison of PUF Parameters and Areas for Practical Identification Requirements

The required identification performance of an identification system is determined by its application, but for most practical applications FAR and FRR both $\leq 10^{-6}$, and hence EER $\leq 10^{-6}$, is minimally desired. For many applications, EER even needs to be considerably smaller, e.g. EER $\leq 10^{-9}$ or even EER $\leq 10^{-12}$ can be required for critical systems. Aiming for a lower EER also provides more freedom in selecting an optimal FAR-vs-FRR trade-off. The results from Fig. 5.3 show that with a 64-bit PUF response, only the pairwise-comparison ring oscillator PUF achieves EER $\leq 10^{-6}$. In order to obtain a better identification performance based on the same PUF, longer responses need to be considered as identifiers. In the following we examine which response lengths, denoted as n_{id}, are needed for every considered PUF to reach an identification performance of respectively EER $\leq 10^{-6}$, EER $\leq 10^{-9}$ and EER $\leq 10^{-12}$.

To ultimately obtain an objective comparison of the identification performance of the different considered PUFs, we also need to take their silicon area efficiency into account. After determining the minimal required response length n_{id} to achieve a certain identification performance, we estimate the required silicon area a PUF construction needs to occupy in order to produce a response of that length. These area estimations are based on the area breakdown of our test chip as presented in Table 4.1. We use the following approach for scaling the area of the different PUFs:

- For the memory-based PUFs, we scale the area completely bitwise, i.e. the estimated silicon area of the entire PUF block as presented in Table 4.1 is divided by the total number of bit cells of the considered PUF implemented on the test chip and multiplied by n_{id}. This is a very rough estimation as it does not take fixed overhead into account, and it assumes bit cells can be instantiated one by one. Especially for the SRAM PUF this is not very accurate since it neglects the overhead of the address decoder, sense amplifiers, readout circuitry, etc. Moreover, typical SRAM array implementations come in multiples of kilobytes, not bits. However, given the limited knowledge we have about the implemented PUF areas, this is the best estimate we can give. For the other memory-based PUFs the estimate is better since they can be instantiated on a bit by bit basis. For the D flip-flop PUF with the chained read-out implementation, this estimation is even fairly accurate.
- For both arbiter-based PUFs, the silicon area is relatively independent of the required number of response bits, since a single 64-bit arbiter PUF can technically produce 2^{64} response bits. It will become clear that for most applications, large problems arise when too many response bits of a single arbiter-based PUF are used, but for plain identification (no authentication) this is not an issue. This means that the identification performance of a single arbiter PUF is virtually unlimited. The area of a single basic arbiter PUF is estimated by dividing the estimated area of the entire arbiter PUF block as reported in Table 4.1 by the total number of instantiated arbiter PUFs on the test chip. The estimated area of the 2-XOR arbiter PUF is that of two basic arbiter PUFs.
- For both ring oscillator PUFs, responses are bit vectors which are computed from 16 simultaneously measured ring oscillator frequencies. The pairwise comparison method produces 8 bits per response, and the Lehmer-Gray method 49 bits. Multiples of these response lengths are obtained by incrementing the number of oscillators in each of the 16 batches by the same amount. The required area for these ring oscillator PUFs is estimated by dividing n_{id} by 8 and 49 respectively, and multiplying the ceiled outcome of this division with the area of a set of 16 oscillators. The area of a single oscillator is estimated by dividing the total area of the ring oscillator block from Table 4.1 by the number of oscillators implemented on the test chip. This estimation is not very accurate as it neglects the overhead of the frequency counters.

The results of these estimations are presented in Tables 5.1, 5.2 and 5.3, respectively for EER $\leq 10^{-6}$, EER $\leq 10^{-9}$ and EER $\leq 10^{-12}$. We want to point out again that the reported silicon area results need to be considered as rough estimates at best, and even better as indications of order of magnitude.

From Tables 5.1 to 5.3 we conclude that, even though only based on rough estimations, SRAM PUFs exhibit by far the best area efficiency for a required identification performance, being an order of magnitude smaller than the next best PUF. This is due to a combination of their relatively strong PUF behavior, and in particular

Table 5.1 Comparison of PUF parameters and estimated silicon area for an identification system with EER $\leq 10^{-6}$

PUF class	n_{id}	t_{id}	$\log_{10}$ FAR	$\log_{10}$ FRR	Silicon area (μm^2)
SRAM PUF	89	21	−6.00	−6.03	72.3
Latch PUF	9344	2664	−6.00	−6.01	77562.5
D Flip-Flop PUF	448	128	−6.05	−6.16	5359.4
Buskeeper PUF	223	72	−6.03	−6.05	1034.4
Arbiter PUF (basic)	101	23	−6.13	−6.21	1089.8
Arbiter PUF (2-XOR)	142	42	−6.06	−6.12	2179.7
Ring Oscillator PUF (P.C.)	62	12	−6.06	−6.20	7531.3
Ring Oscillator PUF (L.G.)	121	30	−6.12	−6.24	2824.2

Table 5.2 Comparison of PUF parameters and estimated silicon area for an identification system with EER $\leq 10^{-9}$

PUF class	n_{id}	t_{id}	$\log_{10}$ FAR	$\log_{10}$ FRR	Silicon area (μm^2)
SRAM PUF	143	34	−9.11	−9.04	116.2
Latch PUF	14881	4243	−9.00	−9.02	123523.9
D Flip-Flop PUF	703	201	−9.00	−9.09	8409.9
Buskeeper PUF	355	115	−9.02	−9.09	1646.7
Arbiter PUF (basic)	160	37	−9.04	−9.39	1089.8
Arbiter PUF (2-XOR)	226	67	−9.15	−9.09	2179.7
Ring Oscillator PUF (P.C.)	104	20	−9.83	−9.28	12238.3
Ring Oscillator PUF (L.G.)	192	48	−9.13	−9.36	3765.6

Table 5.3 Comparison of PUF parameters and estimated silicon area for an identification system with EER $\leq 10^{-12}$

PUF class	n_{id}	t_{id}	$\log_{10}$ FAR	$\log_{10}$ FRR	Silicon area (μm^2)
SRAM PUF	196	47	−12.0	−12.1	159.3
Latch PUF	20464	5835	−12.0	−12.0	169867.2
D Flip-Flop PUF	968	277	−12.0	−12.1	11580.1
Buskeeper PUF	488	158	−12.1	−12.0	2263.7
Arbiter PUF (basic)	217	50	−12.1	−12.1	1089.8
Arbiter PUF (2-XOR)	312	93	−12.1	−12.3	2179.7
Ring Oscillator PUF (P.C.)	155	29	−14.8	−12.4	18828.1
Ring Oscillator PUF (L.G.)	260	65	−12.1	−12.2	5648.4

their very small cell area. Pairwise comparison ring oscillator PUFs, though showing the overall best identification performance in Fig. 5.3, are very area-inefficient, which makes them almost the least favorable choice for a required performance. Latch PUFs still behave the worst, even when taking into account their silicon area, because they require huge response lengths to provide meaningful levels of identification performance.

5.3 PUF-Based Entity Authentication

5.3.1 Background: PUF Challenge-Response Authentication

When an entity wants to *authenticate* itself to an another party, typically called the *verifier*, it needs to provide, besides plain identification, also corroborative evidence of its presented identity, i.e. evidence which could only have been created by that particular entity. An entity typically achieves this goal by proving to the verifier that it knows, possesses or contains a particular secret which only that entity can know, have or contain. In addition, the entity also needs to convince the verifier that it was actively, i.e. at the time of authenticating, involved in creating that evidence. In this section, we discuss how an entity can authenticate itself based on the possession/containment of a unique PUF instance. An authentication scenario is considered with a centralized verifier and entities which authenticate to the central verifier, not to each other.

There are two main approaches towards developing a PUF-based authentication system. The first approach is to develop an authentication scheme which directly deploys the unique and unpredictable challenge-response behavior of a particular PUF instance. The second approach consists of deriving a robust and secure cryptographic key from a PUF response and using this key in an existing classic key-based cryptographic authentication protocol. PUF-based key generation is discussed in detail in Chap. 6. In this section, we focus on the challenge-response approach, which is usually more efficient, since it does not require an additional implementation of a key generation algorithm and other keyed cryptographic primitives.

Basic PUF-Based Challenge-Response Authentication

The basic PUF-based challenge-response entity authentication scheme, among others described by Gassend et al. [42], Ranasinghe et al. [108] and Devadas et al. [30], consists of two phases, *enrollment* and *verification*:

1. Before deployment, every entity goes through enrollment by the verifier. During the enrollment phase, the verifier records the identity[3] ID of every entity, and

---

[3]Identification can happen with both assigned or inherent identifiers, as discussed in detail in Sect. 5.2. For simplicity, but without loss of generality, we assume an entity identifies with an assigned identifier ID in this section.

collects a significant subset of challenge-response pairs of every entity's PUF, for randomly generated challenges. The collected challenge-response pairs are stored in the verifier's database DB, indexed by the entity's ID.

2. During the verification phase, an entity identifies itself to the verifier by sending its ID. The verifier looks up the ID in its DB, and selects a random PUF challenge-response pair stored with that ID. The PUF challenge is sent to the entity; the entity evaluates its PUF with that challenge and replies with the obtained response. The verifier checks whether the replied response is *close to* the response it has in its database, i.e. both responses differ no more than some predetermined authentication threshold t_{auth}. If this check succeeds, the entity is authenticated, otherwise the authentication is rejected. The used challenge-response pair is removed from DB.

The correctness of this authentication scheme is ensured by the fact that PUF responses are reproducible over time, up to a small intra-distance. If t_{auth} is set large enough such that with high probability the intra-distance is smaller, the authentication of a legitimate entity succeeds. For the security of the authentication, the verifier relies on the fact that PUFs are unique and unclonable, and only the genuine PUF can with high probability reproduce a close response to a previously unobserved and random challenge.

Drawbacks of the Basic Protocol

Unfortunately, while strikingly simple and low-cost, this basic protocol exhibits a number of major shortcomings and drawbacks:

- It is clear that challenge-response pairs cannot be reused in order to avoid replay attacks, and a used pair is therefore removed from DB after the protocol finishes. This entails that the verifier needs to store a large number of pairs for each entity to make sure that every entity can be authenticated a reasonable number of times. Maintaining such a large database is a significant effort. Moreover, it requires that the considered PUF construction has a large challenge set to begin with, which already rules out a significant number of proposed PUF constructions.
- When the stored challenge-response pairs in DB of an entity run out, the entity can no longer be authenticated by the verifier. To make further authentications possible, the entity needs to be re-enrolled. This requires either a physical withdrawal of the entity from the field to undergo re-enrollment at the verifier's secured premises, or an elaborate and costly extension of the basic protocol to make remote re-enrollment possible.
- The basic protocol only provides authentication of an entity to the verifier; no mutual authentication can be supported without significantly extending the basic protocol.
- The basic protocol is only secure for truly unclonable PUFs (cf. Definition 13), i.e. PUFs which are also mathematically unclonable. PUFs which can be cloned in a mathematical sense (and which are hence still considered PUFs according to

Definition 16) do not offer secure authentication, since the basic protocol can no longer distinguish between the real entity with the physical PUF and an impersonator with a mathematical clone of that PUF. From Table 3.1, it is clear that only the optical PUF presents convincing evidence to be considered mathematically unclonable, which means the basic scheme is currently only secure when an optical PUF is deployed.[4]

Improvements and extensions of this basic protocol have been proposed, e.g. by Bolotnyy and Robins [9] and Kulseng et al. [71], to relax the strict requirements on the PUF construction and the database. However, these proposals do not completely succeed in overcoming the shortcomings of the basic protocol, or they introduce new restrictions. Some of these proposals are moreover shown to be insecure, e.g. Kardas et al. [63] point out significant security weaknesses of the proposal by Kulseng et al. [71].

Motivation

The simplicity of the basic PUF-based challenge-response authentication protocol is very appealing, especially when entities are heavily constrained silicon devices such as RFID tags, which cannot dedicate many resources to implementations of cryptographic building blocks. A possibly significant improvement in the resource-security trade-off of such devices is possible, if they can be authenticated only based on an efficient embedded intrinsic PUF implementation. Unfortunately, no intrinsic PUF candidates currently exist which meet the very strong requirement of mathematical unclonability needed to make the basic challenge-response authentication protocol secure. Constructing a practical intrinsic PUF for which strong guarantees of mathematical unclonability can be provided is currently considered an important open problem in the study of PUFs and their constructions.

Instead of attempting to tackle this difficult open problem, we will propose an alternative PUF-based authentication protocol in this section, which has considerably relaxed requirements for the PUF used, and which moreover provides mutual authentication. We need a little more complexity than the basic protocol (less is virtually impossible), but aim to keep it at a minimum, especially on the entity's side, which we consider to be a resource-constrained device like an RFID tag.

5.3.2 A PUF-Based Mutual Authentication Scheme

Rationale

The premise on which we base our proposed protocol is that the amount of unpredictability in the challenge-response behavior of an intrinsic PUF instance is strictly

---

[4]Intrinsic PUF constructions have been proposed which are candidates for mathematical unclonability, but currently they lack convincing argumentation to be classified as such.

limited, and increasing it comes at a high relative implementation cost. Note that this is the case for all currently known intrinsic PUF constructions. It is therefore not recommendable to publicly disclose challenge-response pairs in the protocol's communications, as the basic protocol does, since every disclosed pair significantly reduces the remaining unpredictability of the PUF's behavior. This quickly results in a situation where the PUF-carrying entity can no longer be securely authenticated.

In classic cryptographic challenge-response authentication protocols based on symmetric-key techniques (cf. [96, Sect. 10.3.2]), an entity also does not authenticate itself by disclosing its secret key, since this would render the authentication insecure after one protocol run. Instead, an entity only proves its knowledge of the secret key by showing that it can calculate encryptions or keyed one-way function evaluations of randomly applied challenges, without revealing any information about the key's value. We follow the same line of thought in our protocol: an entity demonstrates its possession of a PUF instance by demonstrating that it can calculate a function evaluation which takes an unpredictable PUF response as input, without fully disclosing the unpredictable nature of the response in the result of this evaluation.

An obvious choice for such a concealing function would be a one-way function, e.g. instantiated as a cryptographic hash function, since a one-way function evaluation of a PUF response does not disclose any significant information about the response value. The verifier could then calculate the same one-way function evaluation on the response in its database and check if this matches the entity's response. However, this runs into the problem of the non-perfect reproducibility of the PUF's responses. Whereas the entity's PUF response value will typically be *close to* the enrolled PUF response value in the verifier's database, the one-way function evaluations of both are completely independent and cannot be meaningfully matched to each other. This is also a consequence (in this case negative) of the one-way function destroying any predictability between input and output.

To overcome this issue, we need to deploy some form of error-correction in the protocol to make sure that both parties, the entity and the verifier, compute the one-way function on exactly the same PUF response value. In order to be able to do this, the entity and the verifier need to publicly exchange an amount of information about the PUF response. This is typically called *side information*, or also *helper data* in the context of PUF-based key generation algorithms (cf. Chap. 6). We show that it is possible for some intrinsic PUFs to find a good balance between the amount of side information one needs to disclose and the amount of unpredictability that is left in the PUF response after observing the side information, to obtain a reliable but secure authentication. Note that at no point in our proposed protocol do we derive a cryptographically secure key or do we use a keyed cryptographic primitive. Moreover, while being related to the key-generation algorithms discussed in Chap. 6, the used error-correction technique in this protocol is considerably less computationally expensive, especially on the entity's side.

Background   The mutual authentication scheme as described in [151], of which a slightly adapted version is presented in this section, is the shared result of numerous fruitful discussions and an intense research collaboration between the author and Anthony van Herrewege, Roel Peeters and Prof. Ingrid Verbauwhede (all of the University of Leuven), and Christian Wachsmann, Prof. Stefan Katzenbeisser and Prof. Ahmad-Reza Sadeghi (all of the Technische Universität Darmstadt). The key idea and development of the "reverse" secure sketch is due to the author.

"Reversed" Secure Sketching Based on Linear Block Code Syndromes

The error-correction technique we deploy in the protocol is a practical version of the syndrome construction of a secure sketch, as described by Dodis et al. [32, 33], with the sketching procedure executed by the entity, and the recovery procedure executed by the verifier. Note that this execution order implies that the entity generates the "correct" version of the PUF response and the verifier needs to correct his stored response value to match that of the entity. This is the opposite way, as secure sketching is typically deployed, e.g. in PUF-based key generation, with the verifier storing a single fixed secret key and helper data string and requiring the entity to correct its noisy response, with the help of the helper data, to generate the same key as the verifier. For this reason, we call this a *reverse secure sketch*. The use of such a reverse secure sketch has some peculiar and interesting side effects:

- Since the entity's PUF response is only non-perfectly reproducible, the actual response value on which the authentication is based will possibly be different for each run of the protocol. This means the exchanged side information will also be different for each run. Since the side information partially discloses the response value, we need to consider this when showing the security of the protocol, i.e. we do not want multiple runs of the protocol to disclose the full response value. We are able to prove that even after an indeterminate number of protocol runs, with each a possibly different side information string based on a different noisy response value, the remaining unpredictability of the PUF response remains high.
- Since the sketching procedure of a secure sketch is typically much less computationally complex than the recovery procedure, it can be efficiently implemented by each entity using a small amount of resources. The central verifier is assumed to be less resource-constrained and easily capable of implementing the recovery procedure.

Secure sketch constructions based on linear block codes are explained in detail in Sect. 6.2.1. We refer to that section for more background on the operation of the syndrome construction used in our protocol. Summarized very briefly:

- The sketching procedure of the syndrome construction, executed by an entity, consists of a binary matrix multiplication of the entity's PUF response evaluation with the parity-check matrix of a linear block code, resulting in the side information for the protocol run: $w_i := y_i' \cdot \mathbf{H}^\mathsf{T}$.
- The recovery procedure, as executed by the verifier, consists of three steps: (i) a syndrome is calculated based on the received side information and the stored PUF response value from the verifier's database: $s_i := w_i \oplus y_i \cdot \mathbf{H}^\mathsf{T} \equiv e_i \cdot \mathbf{H}^\mathsf{T}$, with

Entity Ent_i : ID_i, puf_i		Verifier Ver : DB
$\text{puf}_i \rightarrow y_i'$		
$w_i := y_i' \cdot \mathbf{H}^\mathsf{T}$		
$r_1 \xleftarrow{\$} \{0,1\}^\ell$	$\xrightarrow{\quad \text{ID}_i, w_i, r_1 \quad}$	Identify ID_i: $\text{DB}[\text{ID}_i] \rightarrow y_i$
		$y_i'' := \text{Recover}(y_i, w_i)$
		$r_2 \xleftarrow{\$} \{0,1\}^\ell$
$\text{Hash}(\text{ID}_i, w_i, y_i', r_1, r_2) \overset{?}{=} u_1$	$\xleftarrow{\quad u_1, r_2 \quad}$	$u_1 := \text{Hash}(\text{ID}_i, w_i, y_i'', r_1, r_2)$
· No match $\rightarrow$ Abort		
· Match $\rightarrow$ Accept Ver		
$u_2 := \text{Hash}(\text{ID}_i, y_i', r_2)$	$\xrightarrow{\quad u_2 \quad}$	$\text{Hash}(\text{ID}_i, y_i'', r_2) \overset{?}{=} u_2$
		· No match $\rightarrow$ Abort
		· Match $\rightarrow$ Accept Ent_i

Fig. 5.4 A PUF-based mutual authentication scheme between a PUF-carrying entity and a central verifier

$e_i = (y_i' \oplus y_i)$; (ii) the syndrome is decoded to an error string: $\text{Decode}(s_i) \rightarrow e_i'$, with $e_i' = e_i$ if $\mathbf{HD}(y_i; y_i') \leq t_{\text{auth}}$, with t_{auth} the bit error correction capacity of the underlying code; (iii) the response reconstruction: $y_i'' := y_i \oplus e_i'$, with $y_i'' = y_i'$ if $\mathbf{HD}(y_i; y_i') \leq t_{\text{auth}}$.

The Mutual Authentication Protocol

The execution flow of our proposed PUF-based mutual authentication protocol is shown in Fig. 5.4. This is a modified version of the protocol we have introduced in [151], the main difference being the reversal of the authentication checks such that an entity will only authenticate to a legitimate verifier. We explain the different operations in more detail:

- Each entity (Ent_i) is assigned a unique identifier ID and equipped with a unique PUF instance puf_i. We only require a single challenge-response pair per PUF; hence we do not explicitly write the challenge: $\text{puf}_i \rightarrow y_i'$.
- Prior to deployment, all entities are enrolled by the verifier (Ver), which keeps a database DB of a single response evaluation y_i of every entity's PUF, indexed by the entity's ID. PUF responses can only be enrolled once. This is enforced by physically blocking or even destroying the entity's enrollment interface which directly outputs the PUF response.
- Each entity implements the sketching procedure of the secure sketch, which is simply a binary matrix multiplication with a parity-check matrix $\mathbf{H}^\mathsf{T}$ of a linear block code. Due to the special structure of these matrices for many block codes, this multiplication can often be very efficiently implemented in digital hardware.
- The verifier implements the recovery procedure of the secure sketch: $\text{Recover}(y_i, w_i)$. This procedure contains an error-correction decoding algorithm which typically requires a considerable computational effort.

- Each entity, as well as the verifier, implements a cryptographically secure hash function: $\mathsf{Hash}(\cdot)$.
- Each entity, as well as the verifier, has access to a cryptographically secure random bit generator which they use to generate random nonces used in the protocol: $r_i \leftarrow \{0, 1\}^\ell$. The nonces are used to introduce freshness in the protocol's messages in order to avoid replay attacks. Later we propose a possible optimization in which entities do not need to produce a nonce and hence do not require a random bit generator.

Correctness

The correctness of the protocol is guaranteed by the error-correction capability of the underlying linear block code of the secure sketch. If, with high probability, the intra-distance between the PUF response produced by a legitimate entity during a protocol run and the response in the legitimate verifier's database is smaller than or equal to the error-correcting capacity t_{auth} of the underlying block code, then the verifier is able to recover the same response value produced by the entity. In that case, both hash value checks will succeed and mutual authentication is accomplished. Based on the intra-distance distribution of the deployed PUF construction, a code with an appropriate error-correction threshold is selected. This will ultimately result in a false rejection rate, in a manner equivalent to PUF-based identification as discussed in Sect. 5.2.2.

Security

Next, we discuss the different security aspects of the protocol. We refer to [151] for a detailed security analysis, including a security proof, of a variant of the presented protocol.

Physical Unclonability   Due to the physical unclonability of the deployed PUFs, an impersonation attempt of an entity carrying a different PUF will fail with high probability. Equivalently to the identification system discussed in Sect. 5.2.2, a false acceptance rate can be computed which depends on the inter-distance distribution of the PUFs, the number of bits n_{auth} in the considered PUF responses, and the selected error-correction threshold t_{auth}. Note that this false acceptance rate expresses the probability that two different PUFs are, coincidentally, similar enough to impersonate each other. The actual false acceptance rate of the overall protocol also depends on the collision resistance of the used hash function and could be considerably higher.

Replay Attacks   The random nonces r_1 and r_2 respectively generated by the entity and the verifier preclude replay attacks from both sides, by introducing freshness in the protocol communications. An adversary trying to replay protocol messages in order to impersonate the verifier will fail since the entity will present a different nonce value and the adversary cannot recompute the hash evaluation over this new nonce since he doesn't know the PUF response. The same holds for an adversary trying to impersonate an entity by replaying earlier recorded messages from a successful entity authentication. Under certain conditions, the nonce generated by the entity can even be omitted. This is the case if the expected intra-distance on the entity's PUF response is *large* enough such that it is highly unlikely that exactly the same response value will ever be reproduced. In that case, the freshness of the protocol is guaranteed by the *noise* on the entity response, given that it is large enough. A practical advantage of this variant is that entities do not need access to a random bit generator any longer, which reduces the resource requirements of the protocol.

Response Unpredictability   We still need to assess the unpredictability of responses, given that the adversary can observe multiple side information strings from many successful authentication attempts of an entity. We first consider the unpredictability of a response after observing one side information string. After observing a protocol run, an adversary can launch an offline attack on the unknown response value y_i' based on the observed quartet $(\mathsf{ID}_i, r_1, w_i, u_1)$, by guessing a response value y_i^* and checking whether $\mathsf{Hash}(\mathsf{ID}_i, w_i, y_i^*, r_1) \stackrel{?}{=} u_1$. Note that, by observing the side information, an adversary gains quite some information about the response value y_i', since every bit of w_i is a linear combination of bits from y_i', which helps it in the guessing attack. We express the remaining unpredictability of the response as its conditional entropy when conditioned on the side information. It is rather trivial to show that $H(Y_i'|W_i) \geq H(Y_i') - |W_i|$. This means that to ensure there is any unpredictability left, the length of the side information needs to be smaller than the response's entropy. Now we show that the unpredictability of the PUF response remains this high even after observing many side information strings computed over different (possibly noisy) evaluations of the same response, by proving the following lemma:[5]

Lemma 1 *If* $\forall j$: Y_i' *and* D_j^{intra} *are pairwise independently distributed, then*:

$$H\left(Y_i'|f\left(Y_i'\right), f\left(Y_i' \oplus D_1^{\mathsf{intra}}\right), f\left(Y_i' \oplus D_2^{\mathsf{intra}}\right), \ldots, f\left(Y_i' \oplus D_q^{\mathsf{intra}}\right)\right) = H\left(Y_i'|f\left(Y_i'\right)\right),$$

for any positive integer q *and for any linear function* f.

Proof

$$H\left(Y_i'|f\left(Y_i'\right), f\left(Y_i' \oplus D_1^{\mathsf{intra}}\right), f\left(Y_i' \oplus D_2^{\mathsf{intra}}\right), \ldots, f\left(Y_i' \oplus D_q^{\mathsf{intra}}\right)\right),$$

    (f is a linear function),

$$= H\left(Y_i'|f\left(Y_i'\right), f\left(Y_i'\right) \oplus f\left(D_1^{\mathsf{intra}}\right), f\left(Y_i'\right) \oplus f\left(D_2^{\mathsf{intra}}\right), \ldots, f\left(Y_i'\right) \oplus f\left(D_q^{\mathsf{intra}}\right)\right),$$

---

[5]Note that this proof differs from the security proof given in [151].

$$= H\big(Y_i'|f\big(Y_i'\big), f\big(D_1^{\text{intra}}\big), f\big(D_2^{\text{intra}}\big), \ldots, f\big(D_q^{\text{intra}}\big)\big),$$

$$\big(\forall j : Y_i' \text{ and } D_j^{\text{intra}} \text{ are independent}\big),$$

$$= H\big(Y_i'|f\big(Y_i'\big)\big). \qquad\qquad \square$$

The assumption of independence between the distributions of response values and intra-distances is very reasonable since they originate from different physical processes. This lemma is directly applicable to our situation, since every helper data string is a linear function of a response evaluation. Based on a similar reasoning, one can also show that the remaining *min-entropy* remains high. This is proven in a generalized form by Boyen [13]. The response's unpredictability as expressed by $H(Y_i'|W_i)$ measures the resistance against an offline attack; hence it needs to be sufficiently large to offer long-term security.

5.3.3   Authentication Performance of Different Intrinsic PUFs

Proposed Entity Design

We first determine a concrete and realistic design of an entity which is used to compare the authentication performance of different intrinsic PUFs in the proposed protocol. The main design choice is the selection of the underlying error-correcting linear block code of the secure sketch. We refer to Burr [18, Chap. 6] for a detailed introduction to block codes. We propose using a concatenation of a simple repetition code followed by a BCH code as introduced by Hocquenghem [51] and Bose and Ray-Chaudhuri [12], which yields an overall syndrome construction with a relatively high error correction performance. The BCH code's parameters are $[n_{\text{BCH}}, k_{\text{BCH}}, t_{\text{BCH}}]$, which means that it has code words of length n_{BCH} and dimension k_{BCH}, and up to t_{BCH} bit errors in a single code word can be corrected. The corresponding parameters of the repetition code are $[n_{\text{REP}}, 1, \frac{n_{\text{REP}}-1}{2}]$, with n_{REP} odd.

An $[n, k, t]$ binary linear block code can be fully described by its generation matrix $\mathbf{G}_{k \times n}$ or its corresponding parity-check matrix $\mathbf{H}_{(n-k) \times n}$, which meet the condition $\mathbf{G} \cdot \mathbf{H}^{\mathsf{T}} = \mathbf{0}$. For a repetition code, a multiplication of an n_{REP}-bit word with the code's parity-check matrix can be implemented straightforwardly in hardware using $(n_{\text{REP}} - 1)$ 2-input XOR gates. Due to its special algebraic structure, the multiplication of an n_{BCH}-bit word with the parity-check matrix of a BCH code can also be implemented very efficiently using an $(n_{\text{BCH}} - k_{\text{BCH}})$-bit linear feedback shift register (LFSR) with the feedback taps determined by the generator polynomial of the BCH code. The overall side information generation of a concatenated repetition and BCH code can hence be efficiently implemented using only minimal resources on the entity's side. A schematic representation of the side information generator is shown in Fig. 5.5. Note that in the proposed design, the BCH code's side information and the verifier authentication hash value are only computed over

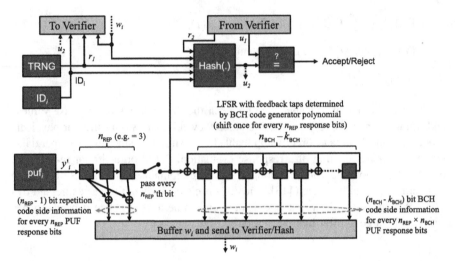

Fig. 5.5 Design of an entity's side information generator, based on a concatenated repetition and BCH block code, and the corresponding verifier authentication check

every n_{REP}'th response bit. Since the remaining $n_{REP} - 1$ bits are immediately dis-closed by the repetition code's side information, it makes no sense to consider them further for authentication and they are discarded.

For the hash function implementation, we propose using a lightweight hash function. In our prototype implementation in [151] we used the SPONGENT hash function as proposed by Bogdanov et al. [8]. To keep the side information gen-erator small, we constrain the repetition code to $n_{REP} \leq 11$ and the BCH code to $n_{BCH} \leq 255$. We now investigate the implementation parameters of this proposed design, within these constraints, when the deployed PUF is one of the intrinsic PUFs studied in Chap. 4. In particular, we determine the required code parameters based on the intrinsic PUFs' statistics summarized in Table 4.10. Based on these code parameters, the required number of PUF response bits, and ultimately the required PUF silicon area to reach a particular authentication performance, can be calculated.

Example of Authentication Performance Calculation

We demonstrate how the authentication performance is calculated for a single PUF in an exemplary but realistic scenario. Next, we present the results of the same calculations for all the intrinsic PUFs studied in a number of possible scenarios.

As an example, we consider the SRAM PUF, and we aim for an authentica-tion performance with EER $\leq 10^{-9}$ and an unpredictability of at least 128 bits, i.e. $H(Y|W) \geq 128$. Searching the design space constrained by the proposed entity design above yields the following parameters:

- A $[n_{REP} = 3, 1, 1]$ repetition code and a $[n_{BCH} = 223, k_{BCH} = 83, t_{BCH} = 21]$ BCH code are used. In total, $3 \times$ a BCH code length is required.

- These code parameters achieve an overall FRR $= 10^{-9.63}$ and FAR $= 10^{-104.84}$.
- The total number of required PUF response bits is $n_{auth} = 3 \times n_{REP} \times n_{BCH} = 2007$.
- The total number of exchanged side information bits is $\ell_{auth} = 3 \times (n_{REP} \times n_{BCH} - k_{BCH}) = 1785$.
- The remaining entropy is calculated as $H(Y|W) \geq H(Y) - |W| = n_{auth} \times \rho(Y^{n_{auth}}) - \ell_{auth} = 2007 \times 94.09\% - 1758 = 130.4$ bits.

Note that these parameters are optimized within the given constraints to yield to smallest possible number of required PUF bits. For most intrinsic PUFs, the entropy density $\rho(Y^{n_{auth}})$ is a constant and independent of the number of considered response bits, as shown in Table 4.10. However, for both arbiter-based PUFs, $\rho(Y^{n_{auth}})$ is a decreasing function of the required number of bits n_{auth}, of which an upper bound is given by Fig. 4.5.

Performance Comparison of Different Intrinsic PUFs

Now we apply the same calculation as in this example on all studied intrinsic PUFs for three different authentication performances:

1. A low-cost security scenario with EER $\leq 10^{-6}$ and $H(Y|W) \geq 80$.
2. A realistic security scenario with EER $\leq 10^{-9}$ and $H(Y|W) \geq 128$.
3. A critical security scenario with EER $\leq 10^{-12}$ and $H(Y|W) \geq 256$.

Based on the number of response bits needed, the required silicon area is also calculated in the same manner (and with the same disclaimers) as in Sect. 5.2.3. The results are presented in Table 5.4.

From Table 5.4, it is clear that for many studied intrinsic PUFs, no parameter solutions meeting the presented design constraints can be found. This is mostly a result of a too high average intra-distance, a too low entropy density, or a combination of both. Only the SRAM PUF and the pairwise comparison ring oscillator PUF succeed in providing a solution for all three considered scenarios. The SRAM PUF solution offers a better area efficiency than the ring oscillator PUF by almost two orders of magnitude.

5.4 Conclusion

In Sect. 5.2 of this chapter, we have successfully demonstrated that a PUF response can be used as a secure and reliable inherent identifier of a PUF-embedding entity. The identification performance, in terms of false acceptance, false rejection and equal error rate, scales only with the bit length of the PUF response. For the intrinsic PUFs studied in Chap. 4, this ultimately comes down to a scaling with the required silicon area needed to implement the PUF. The scaling factor is different for every intrinsic PUF construction and is determined by the characteristics of its inter-

Table 5.4 Required code parameters, PUF response bits and estimated silicon area for achieving three different authentication performances. '/' means that no parameter solution within the given design constraints can be found

PUF class	×BCH	n_{REP}	n_{BCH}	k_{BCH}	t_{BCH}	$\log_{10}$ EER	$H(Y'|W)$	n_{auth}	Silicon area (μm^2)
EER $\leq 10^{-6}$ and $H(Y'|W) \geq 80$									
SRAM PUF	1	3	248	124	18	−7.05	80	744	604.5
Latch PUF	/	/	/	/	/	/	/	/	/
D Flip-Flop PUF	/	/	/	/	/	/	/	/	/
Buskeeper PUF	/	/	/	/	/	/	/	/	/
Arbiter PUF (basic)	/	/	/	/	/	/	/	/	/
Arbiter PUF (2-XOR)	1	5	232	100	19	−6.2	80	1160	2179.7
Ring Oscillator PUF (P.C.)	2	1	216	52	25	−6.1	81	432	50835.9
Ring Oscillator PUF (L.G.)	/	/	/	/	/	/	/	/	/
EER $\leq 10^{-9}$ and $H(Y'|W) \geq 128$									
SRAM PUF	3	3	223	83	21	−9.63	130	2007	1630.7
Arbiter PUF (2-XOR)	/	/	/	/	/	/	/	/	/
Ring Oscillator PUF (P.C.)	2	3	167	91	10	−9.19	128	1002	117911.1
EER $\leq 10^{-12}$ and $H(Y'|W) \geq 256$									
SRAM PUF	7	3	249	81	26	−12.37	257	5229	4248.7
Ring Oscillator PUF (P.C.)	8	1	254	46	42	−14.35	260	2032	239117.2

and intra-distance distributions. A comparative analysis, as presented in Tables 5.2 and 5.3, indicates that for a given silicon area, the SRAM PUF provides the best identification performance in terms of equal error rate. The latch PUF on the other hand exhibits very poor identification capabilities, which severely undermines its usage as a PUF.

We build upon these identification capabilities of intrinsic PUFs to provide secure entity authentication. In order to obtain a cryptographic level of authentication security, we proposed a new authentication protocol between a central verifier and a deployed entity, shown in Fig. 5.4, that utilizes a unique and unpredictable response of a PUF instance embedded by each entity as an authentication secret. The protocol requires some form of error-correction which we accomplish by using a secure sketch in a 'reversed' mode of operation, i.e. an entity generates new side information in each protocol run and the verifier uses this side information to modify its fixed PUF response to match the current response evaluation of the entity. Based on the linearity of the sketching procedure, we are able to prove that even after many protocol runs, the unpredictability of the PUF response remains high. The authentication performance metrics of this protocol are derived in a similar way as for the PUF-based identification, and applied on the experimental intrinsic PUF results. From this analysis, summarized in Table 5.4, it is clear that most of the intrinsic PUFs studied in Chap. 4 are not able to achieve a high-level authentication performance within a resource-constrained environment. The SRAM PUF and the pairwise-comparison ring oscillator PUF, and to a lesser extent also the 2-XOR arbiter PUF, are the only constructions which are sufficiently unpredictable and reproducible to be practically usable in the protocol, with the SRAM PUF exhibiting the best silicon area efficiency.

Chapter 6
PUF-Based Key Generation

6.1 Introduction

6.1.1 Motivation

An indispensable premise for a large majority of cryptographic implementations
is the ability to securely generate, store and retrieve keys. While often considered
trivial in the description of cryptographic primitives, the required effort to meet
these conditions is not to be underestimated. The minimal common requirements
for secure key generation and storage are: (i) a source of randomness to ensure that
freshly generated keys are unpredictable and unique, and (ii) a protected memory
which reliably stores the key's information while shielding it completely from unauthorized parties. From an implementation perspective, both requisites are non-trivial
to achieve. The need for unpredictable and unique randomness is typically filled by
applying a (seeded pseudo-)random bit generator (PRNG). However, the fact that
such generators are difficult to implement properly was just recently made clear
again by the observation by Lenstra et al. [77] that a large collection of "random"
public RSA keys contains many pairs which share a prime factor, which is immediately exploitable and would not occur if they had been generated based on true
randomness. On the other hand, implementing a protected memory is also a considerable design challenge, often leading to an increased implementation overhead
and/or restricted application possibilities in order to enforce the physical security of
the stored key. Countless examples can be provided of broken cryptosystems due to
poorly designed or implemented key storages, or bad handling of keys. Moreover,
even high-level physical protection mechanisms are often not sufficient to prevent
well-equipped and motivated adversaries from discovering digitally stored secrets,
e.g. as demonstrated by Tarnovsky [138], Torrance and James [143].

A PUF-based key generator tries to tackle both requirements at once by using
a PUF to harvest static but device-unique randomness and processing it into a cryptographic key. This avoids the need for both a PRNG, since randomness which is
already intrinsically present in the device is used, and the need for a protected non-
volatile memory, since the used randomness is considered static over the lifetime of

R. Maes, *Physically Unclonable Functions*, DOI 10.1007/978-3-642-41395-7_6,
© Springer-Verlag Berlin Heidelberg 2013

the device and can be measured again and again to regenerate the same key from the otherwise illegible random features. Due to these very interesting properties, cryptographic key generation is one of the main applications of (intrinsic) PUFs and various PUF-based key generation techniques have been proposed by numerous authors, among others by Bösch et al. [11], Guajardo et al. [45], Tuyls and Batina [144], Yu et al. [158]. Since PUF responses are generally not perfectly reproducible and not uniformly distributed, they can evidently not be used directly as cryptographic keys. A PUF-based key generator faces two main challenges: increasing the reliability to a practically acceptable level and compressing sufficient entropy in a fixed length key.

6.1.2 Chapter Goals

Developing a PUF-based key generator is a process involving many design decisions and trade-offs, of both a pragmatic nature (e.g. information-theoretical versus computational security) and a quantitative nature (e.g. key reliability versus required silicon area). The main goal of this chapter is to investigate the available techniques and methods for generating a usable cryptographic key from a PUF's response, and apply the obtained insight to develop practical PUF-based key generation methodologies, building blocks and systems. In more detail:

- We will study existing notions and constructions used for generating cryptographic keys from entropic sources.
- We will propose improvements and extensions for some of these existing constructions.
- We aim to develop a methodology for *practical* PUF-based key generation by adopting an as realistic as possible model, both for the randomness source and for the envisioned application requirements, and explicitly stating the security design constraints for a given efficient construction in this model.
- Finally, we plan to apply this methodology, first to provide an assessment of the performance and efficiency of the intrinsic PUFs tested in Chap. 4 with regard to key generation, and then to develop a complete and practical PUF-based key generation system.

6.1.3 Chapter Overview

In Sect. 6.2, we give an overview of existing notions and constructions which are of interest for PUF-based key generation, such as secure sketches, strong and fuzzy extractors and entropy accumulators. In Sect. 6.3, we present the main results of our work in [85, 86], which lead to a significantly improved secure sketch implementation based on the use of soft-decision information from the PUF response bits. A complete practical PUF-based key generation system is discussed in

Sect. 6.4. The main contribution of this last section is the presented transition to a more efficient *practical* PUF-based key generation methodology by abandoning the information-theoretical security requirement. In this new mindset, we provide a comparative overview of the key generation performance of the intrinsic PUFs studied in Chap. 4 and finally develop and implement a complete yet efficient and flexible, practical PUF-based key generator. Section 6.5 concludes this chapter.

6.2 Preliminaries

6.2.1 Secure Sketching

Concept

Dodis et al. [32, 33] introduce the concept of a secure sketch as:

Definition 26 A $(\mathcal{Y}, m, m', t)$-secure sketch is a pair of randomized procedures, Sketch and Recover, with the following properties:

- The sketching procedure Sketch on input $y \in \mathcal{Y}$ returns a bit string $w \in \{0, 1\}^*$.
- The recovery procedure Recover takes an element $y' \in \mathcal{Y}$ and a bit string $w \in \{0, 1\}^*$, and returns a bit string $y'' \in \mathcal{Y}$.
- The *correctness property* of secure sketches guarantees that if $\mathbf{dist}[y; y'] \leq t$, then Recover($y'$, Sketch($y$)) = y. If $\mathbf{dist}[y; y'] > t$, then no guarantee is provided about the output of Recover.
- The *security property* guarantees that if y is randomly selected from $\mathcal{Y}$ according to a distribution with min-entropy m, then the value of y can be recovered by the adversary who observes w with probability no greater than $2^{-m'}$. That is, $\widetilde{H}_\infty(Y|\text{Sketch}(Y)) \geq m'$.

A secure sketch is efficient if Sketch and Recover run in expected polynomial time.

The key notion of a secure sketch is that a 'noisy' version y' of an earlier observed value y can be exactly recovered by a recovery procedure when some side information w about the original value is available, without having this side information reveal the complete unpredictability of the value of y. Dodis et al. [33] also propose two practical constructions of secure sketches for a Hamming distance metric. Both constructions are based on the use of a binary error-correcting linear block code. Let $\mathcal{C}$ be a binary error-correcting block code with parameters $[n, k, t]$, i.e. $\mathcal{C}$ contains 2^k different codewords of length n bits which are each at least $2t - 1$ bits apart, and an efficient correction procedure exists which can correct up to t bit errors on each code word. The code is completely characterized by its generator matrix $\mathbf{G_{k \times n}}$ and its parity-check matrix $\mathbf{H_{(n-k) \times n}}$, with $\mathbf{G} \cdot \mathbf{H}^{\mathsf{T}} = \mathbf{0}$. This code can be used to construct an $(\{0, 1\}^n, m, m - (n - k), t)$-secure sketch, as demonstrated by the following two constructions.

The Code-Offset Construction

- Sketch$(y) \rightarrow w$: the sketching procedure samples uniformly at random a code-word $c \overset{\$}{\leftarrow} C$ (independently from the value of y) and the n-bit side information string is the binary offset between y and c: $w := y \oplus c$.
- Recover$(y', w) \rightarrow y''$: the recovery procedure calculates a noisy codeword as $c' := y' \oplus w \equiv (y' \oplus y) \oplus c$ and applies the error-correcting procedure of the code C to correct c': $c'' := $ Correct(c'). The recovered value is computed as $y'' := w \oplus c'' = y \oplus (c \oplus c'')$.
- The correctness property of the construction follows from the error-correction capacity of the underlying code: if $\mathbf{HD}(y; y') \leq t$ then $\mathbf{HD}(c; c') \leq t$ and the correction procedure can perfectly correct the noisy codeword: $c'' = c$, and hence $y'' = y$.
- The security property of the construction is intuitively explained by the fact that w discloses at most n bits of which k ($\leq n$) are independent from y; hence the remaining min-entropy is $\tilde{H}_\infty(Y|W) \geq H_\infty(Y) - (n-k) = m - (n-k)$. For a formal proof we refer to [33].

The Syndrome Construction

- Sketch$(y) \rightarrow w$: the sketching procedure generates an $(n-k)$-bit helper data string as $w := y \cdot \mathbf{H}^\mathsf{T}$.
- Recover$(y', w) \rightarrow y''$: the recovery procedure calculates a syndrome s of the code C as $s := y' \cdot \mathbf{H}^\mathsf{T} \oplus w \equiv (y \oplus y') \cdot \mathbf{H}^\mathsf{T}$. The syndrome decoding procedure of C finds the unique error word e such that $s = e \cdot \mathbf{H}^\mathsf{T}$ and $\mathbf{HW}(e) \leq t$. The recovered value is computed as $y'' := y' \oplus e$.
- The correctness property of the construction follows from the fact that, if $\mathbf{HD}(y; y') \leq t$, then $e = y \oplus y'$ and hence $y'' = y$.
- The security property of the construction follows immediately by application of Lemma 3 from Appendix A.

6.2.2 Randomness Extraction

Concept

In many situations high entropy or nearly uniform bit strings are required, e.g. to serve as a key in key-based security applications. However, often only a low entropy random source is available, e.g. a source which produces heavily biased and/or dependent bits. A method is required to extract near-uniform or 'high quality' randomness from 'low quality' randomness sources with only a limited entropy per bit. This is the purpose of a randomness extractor. We first discuss randomness extraction from a theoretical perspective and next some recommended methods which are often used in practice.

Information-Theoretically Secure Randomness Extraction

The notion of a strong randomness extractor was introduced by Nisan and Zuckerman [101].

Definition 27 Let $\mathsf{Ext} : \{0, 1\}^n \to \{0, 1\}^\ell$ be a polynomial time probabilistic function which uses r bits of randomness. We say that Ext is an efficient (n, m, ℓ, ϵ)-strong extractor if for all random variables $Y \leftarrow \{0, 1\}^n$ with $H_\infty(Y) \geq m$, it holds that

$$\mathsf{SD}\big((\mathsf{Ext}(Y; R), R); (U_\ell, R)\big) \leq \epsilon,$$

where R is uniform on $\{0, 1\}^r$ and U_ℓ is uniform on $\{0, 1\}^\ell$.

An (n, m, ℓ, ϵ)-strong extractor with a small ϵ is hence able to extract high quality (ϵ-uniform) random strings from a low-quality source whose randomness is expressed by its min-entropy.

Randomness extractors can be constructed from so-called universal hash function families as introduced by Carter and Wegman [21]. A hash function family, which is a set of indexed functions $\mathcal{H} = \{h_r : \{0, 1\}^n \to \{0, 1\}^\ell\}_{r \in \mathcal{R}}$, is called universal if

$$\forall a \neq b \in \{0, 1\}^n : \Pr_{r \in \mathcal{R}}\big(h_r(a) = h_r(b)\big) \leq 2^{-\ell}.$$

An important property of universal hash functions is given by the left-over hash lemma, which actually states that a universal hash function is a strong randomness extractor.

Lemma 2 If $\mathcal{H} = \{h_r : \{0, 1\}^n \to \{0, 1\}^\ell\}_{r \in \mathcal{R}}$ is a universal hash function family, then for any random variable $Y \leftarrow \{0, 1\}^n$ it holds that:

$$\mathsf{SD}\big((h_R(Y), R); (U_\ell, R)\big) \leq \frac{1}{2}\sqrt{2^{\ell - H_\infty(Y)}}.$$

From the left-over hash lemma, it follows directly that a universal hash function is a (n, m, ℓ, ϵ)-strong extractor if $\ell \leq m - 2\log_2(\frac{1}{\epsilon}) + 2$. Moreover, from an information-theoretical perspective this is the maximal number of nearly uniform bits which can be extracted from a source with min-entropy m. Observe that one pays a high price in min-entropy to obtain such information-theoretically strong claims. For example, consider that one desires an extracted bit string of length $\ell = 128$ which is $\epsilon = 2^{-128}$-close to uniform, which by itself will have a min-entropy very close to 128 bits. To derive such a string with a strong extractor, one requires a randomness source with a min-entropy of at least $m \geq \ell + 2\log_2 \frac{1}{\epsilon} + 2 = 382$. Hence, at least $382 - 128 = 254$ bits of min-entropy are lost in the extraction process. Moreover, to obtain good randomness extraction, the index r of the used universal hash function (also called the *seed*) needs to be uniform random for every extraction, i.e. one already requires a high-quality random value (of smaller length) to bootstrap the extraction process.

Practical Randomness Extraction

High-quality randomness is often required in practical security applications, e.g. for session keys, random nonces, randomized procedures, etc. However, the earlier described information-theoretic approach is often too restrictive. In many practical systems, entropy collection from truly random physical sources is limited and comes at a rather high cost. It either needs to be collected over relatively long periods of time, or in the case of PUFs, more randomness comes at a direct area cost. The rather wasteful operation of a strong extractor is hence undesirable. It is moreover very difficult or impossible to accurately estimate the (min-)entropy of a randomness source, as required to have strong guarantees for the extractor. This entails that one needs to work with pessimistic underestimations, which causes even more entropy losses.

A more practical approach is considered in the construction of seeded cryptographic pseudo-random number generators. Such PRNGs also take their seed from physical sources of entropy, but abandon the very strict requirement of generating information-theoretically secure randomness. Instead, they are based on a collection of randomness, measured by an estimation of its (Shannon) entropy, and the application of cryptographic primitives to accumulate sufficient entropy in a fixed-length buffer. Through the use of well-studied design principles and after considerable public scrutiny, a number of practical PRNG-constructions are developed which are considered cryptographically secure, and which are widely used in practical applications. These design principles and current best-practice techniques for such constructions are discussed, e.g. by Barker and Kelsey [4], Eastlake et al. [34], Kelsey et al. [65]. We also refer to Gutmann [47, Chap. 6] for an extensive overview of existing PRNG-constructions and a discussion on their (in)security.

The building block of these PRNG-constructions which is of interest to us is the so-called *entropy accumulator*, i.e. the primitive which operates directly on the physical source of randomness in order to produce a highly random seed for the PRNG. Kelsey et al. [65] describe the properties an entropy accumulator should have. Based on this description, we formulate the following informal definition of an entropy accumulator:

Definition 28 An m-bit entropy accumulator is a procedure Acc $: \{0, 1\}^* \to \{0, 1\}^m$ with an internal state of m bits which is called the entropy pool, and which meets the following conditions:

- It is expected that nearly all entropy of the input bitstream is accumulated in the pool, up to the size of the pool, even when the entropy is distributed in various odd ways in this bitstream, e.g. only entropy in every 100th bit, or no entropy in a long stream of bits and then suddenly 100 bits with full entropy, etc.
- An adversary that has momentary control over the input bitstream must not be able to undo the effect of the previous unknown bits which were accumulated in the pool.
- An adversary that has momentary control over the input bitstream must not be able to force the pool into a weak state such that no further entropy can be accumulated.

- An adversary that can choose which bits in the bitstream will be unknown to it, but still has to allow n unknown bits, must not be able to narrow down the number of states of the pool to substantially fewer than 2^{-n}.

When the entropy accumulator assesses that it has accumulated at least m bits of entropy in its entropy pool, the current state of the pool is outputted. This decision can be based on internal (conservative) entropy estimation methods, or on a realistic entropy model of the input bitstream.

In [4, Sect. 10.4], constructions for entropy accumulators based on a generic cryptographic hash function or a block cipher are provided. Kelsey et al. [65] and Ferguson and Schneier [36, Chap. 10] also propose a PRNG design which deploys a cryptographic hash function as an entropy accumulator.

6.2.3 Fuzzy Extractors

Dodis et al. [32, 33] define a fuzzy extractor as:

Definition 29 An $(\mathcal{Y}, m, \ell, t, \epsilon)$-fuzzy extractor is a pair of randomized procedures, Generate and Reproduce, with the following properties:

- The generation procedure Generate on input $y \in \mathcal{Y}$ outputs an extracted string $z \in \{0, 1\}^\ell$ and a helper data string $w \in \{0, 1\}^*$.
- The reproduction procedure Reproduce takes an element $y' \in \mathcal{Y}$ and a bit string $w \in \{0, 1\}^*$ as inputs and outputs an extracted string $z' \in \{0, 1\}^\ell$.
- The correctness property of fuzzy extractors guarantees that if $\mathbf{dist}[y; y'] \leq t$ and z, w were generated by $(z, w) \leftarrow$ Generate(y), then Reproduce$(y', w) = z$. If $\mathbf{dist}[y; y'] > t$, then no guarantee is provided about the output of Reproduce.
- The security property guarantees that if y is randomly sampled from $\mathcal{Y}$ according to a distribution with min-entropy at least m, the extracted string z is nearly uniform even for those who observe w: if $(Z, W) \leftarrow$ Generate(Y), then $\mathbf{SD}((Z, W); (U_\ell, W)) \leq \epsilon$.

A fuzzy extractor is efficient if Generate and Reproduce run in expected polynomial time.

A more generic description of a fuzzy extractor for continuous random variables, in the context of biometrics, was introduced as a *shielding function* by Linnartz and Tuyls [81]. From its definition, it is clear that a fuzzy extractor is a generalization of both a secure sketch, generating a highly random output instead of reconstructing the original input, and a strong extractor, operating on a noisy input instead of a fixed value. Consequently, the basic construction of a fuzzy extractor as proposed by Dodis et al. [32, 33] consists of a straightforward combination of a secure sketch with a randomness extractor. The secure sketch respectively executes its sketching and reconstruction procedures in the fuzzy extractor's generation and reproduction procedures. The strong extractor is the same, and uses the same seed, in both proce-

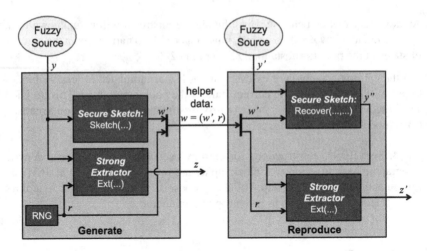

Fig. 6.1 Construction of a fuzzy extractor from a secure sketch and a strong extractor

dures. The side information produced by the secure sketch and the strong extractor's seed, both produced during the generation procedure, are passed to the reproduction procedure as *helper data*. This construction is shown in Fig. 6.1.

Taking two realizations of the same fuzzy secret as an input, this fuzzy extractor construction is able to generate the same (nearly) uniformly random output, i.e. given that the fuzzy source meets the distance and min-entropy conditions of the fuzzy extractor, $z = z'$. Since the full unpredictability of the output is guaranteed even when the helper data is known, the helper data does not need to be kept confidential. This makes it possible to store and transfer the helper data in a public manner. The integrity of the helper data needs to be guaranteed though.

The correctness property of this construction of a $(\mathcal{Y}, m, \ell, t, \epsilon)$-fuzzy extractor is evidently guaranteed by the correctness property of the used $(\mathcal{Y}, m, m', t)$-secure sketch. The security property is taken care of by deploying an (n, m', ℓ, ϵ)-strong extractor, given that $\mathcal{Y} = \{0, 1\}^n$. In other words, the strong-extractor compresses the remaining min-entropy, after publication of the side-information, into a fully random output which can be used as a secret key. Due to the deployment of a randomness extractor, a fuzzy extractor provides the same very high information-theoretic guarantees, but also suffers from the same very stringent min-entropy requirements and losses.

6.3 A Soft-Decision Secure Sketch Construction

6.3.1 Motivation

When deploying a secure sketch to generate a reliable version of a noisy PUF response, it is typically assumed that every single PUF response bit is equally likely to be erroneous in a given evaluation. This is often expressed by a static bit error rate p_e which is estimated as the average ratio of the number of bit errors in the total number of evaluated response bits. Secure sketches are designed to cope with the given bit error rate of the PUF. In particular the error-correcting code parameters, e.g. of the code-offset or syndrome construction, are adapted to this fixed error rate.

In [86], we argue that at least for SRAM PUFs, the assumption of a single fixed error rate is too simplistic and too pessimistic. When analyzing experimental data from an SRAM PUF experiment, it is readily observed that the large majority of the response bits are extremely robust, i.e. they never or hardly ever will change their value between evaluations. A minority of bits are more noisy, occasionally presenting a bit error, and a few bits have no preferred value at all and are basically completely random at each evaluation. This means that every SRAM PUF response bit has its own individual error probability, and the values of these error probabilities are themselves randomly distributed for each bit. In [86] we succeeded in deriving an expression for this distribution based on a simplified model of the construction and operation of an SRAM PUF, and experimentally verified the accuracy of this distribution.

From this observation, it follows that an assumed fixed error probability of p_e for every SRAM PUF response bit is an overestimation for a large majority of bits. A secure sketch designed for such a fixed error rate will hence be overly conservative for most bits. A more efficient secure sketch would take into account the individual likelihood of every response bit being wrong. This is precisely what a secure sketch based on a *soft-decision error correction decoder* could achieve, given that soft-decision information for the response bits is available. We call this a *soft-decision secure sketch*. For an SRAM PUF, soft-decision information for the response bits can be obtained by measuring every bit many times and publishing the observed individual error probabilities as helper data for the soft-decision secure sketch. In [86], we have demonstrated that at least for SRAM PUFs this is a possibility, since the publication of these probabilities does not induce any additional min-entropy loss for the actual unpredictable values of every response bit.

6.3.2 Soft-Decision Error Correction

Soft-decision error correction is a generalization of discrete (hard-decision) error correction. Besides using the redundancy in a codeword to achieve error correction, like a hard-decision decoder, a soft-decision decoder will also consider the estimated reliability of every bit in a codeword and will adapt its correction strategy based on this soft-decision information. As a result, soft-decision decoders are typically more efficient than their hard-decision counterparts, i.e. they can correct more bit errors based on the same amount of redundancy, or equivalently they can achieve the same correction capability by using a code with less redundancy. For the code-offset and

syndrome constructions of secure sketches, based on a binary linear block code
with parameters $[n, k, t]$, the redundancy of $(n - k)$ bits is precisely the amount of
(min-)entropy which is lost due to the publication of the helper data. When a soft-
decision decoder is used, less redundancy is needed and hence less (min-)entropy
is lost. This results in a relaxed requirement on the number of PUF response bits
which are needed to derive a key of a particular length, and ultimately in a smaller
PUF implementation.

   Two well-known soft-decision decoding algorithms are the Viterbi algorithm for
convolution codes, as proposed by Viterbi [152], and the belief propagation algorithm for LDPC codes, as proposed by Gallager [39]. However, the two types are
not really adequate for use in typical secure sketch constructions since they require
very long data streams to work efficiently, while the length of a fuzzy secret such
as a PUF response is often limited. We would like to use a soft-decision decoding
algorithm for rather short linear block codes in order to maintain efficiency. Although less common, soft-decision decoding techniques for short linear block codes
exist.

Soft-Decision Maximum-Likelihood Decoding (SDML)

Soft-decision maximum-likelihood decoding (SDML) is a straightforward algorithm that selects the code word that was most likely transmitted based on the bit re-
liabilities. SDML achieves the best error-correcting performance possible, but generally at a decoding complexity exponential in the code dimension k. Repetition
codes ($k = 1$) can still be efficiently SDML-decoded. Conversely, for $k = n$, SDML
decoding degenerates to making a hard decision on every bit individually only based
on its reliability, and for $k = n - 1$ the block code is a parity-check code and SDML
decoding is done efficiently by flipping the least reliable bit to match the parity. This
last technique is also known as Wagner decoding and is described by Silverman and
Balser [130].

Generalized Multiple Concatenated (GMC) Decoding of Reed-Muller Codes

An r-th order Reed-Muller code $RM(r, m)$ is a binary linear block code with parameters $[n = 2^m, k = \sum_{i=0}^{r} \binom{m}{i}, t = 2^{m-r-1} - 1]$. It is well known that $RM(r, m)$ can
be decomposed into the concatenation of two shorter inner codes, $RM(r - 1, m - 1)$
and $RM(r, m - 1)$, and a simple length-2 block code as outer code. This decompo-
sition can be applied recursively until one reaches $RM(0, m')$, which is a repetition
code, or $RM(r' - 1, r')$ (or $RM(r', r')$), which is a parity check (or degenerated) code,
all of which can be efficiently soft-decision-decoded with SDML. This technique,
introduced as Generalized Multiple Concatenated decoding (GMC) by Schnabl and
Bossert [123], yields a much lower decoding complexity than SDML, but only a
slightly decreased error-correcting capability.

6.3.3 Soft-Decision Secure Sketch Design

Soft-Decision Secure Sketch

In [85], we propose a practical soft-decision secure sketch based on the code-offset construction deploying a concatenated soft-decision decoding of an inner repetition code and an outer Reed-Muller code. Soft-decision information is obtained from response bits during the initial generation procedure and published as helper data. The soft-decision information represents the log-likelihood of every bit being wrong, which is calculated from the estimated error probability $p_e(i)$ of response bit i as:

$$K_i = \left\lfloor \log_\beta \frac{1 - p_e(i)}{p_e(i)} \right\rceil.$$

The log-likelihoods are represented as an 8-bit two's-complement integer value and the logarithm base β is chosen to avoid overflows in this representation.

In the recovery phase of the secure sketch, the regular helper data is combined with the PUF response evaluation to construct a noisy codeword c' as for the regular code-offset construction described in Sect. 6.2.1. The bits of this noisy codeword are now combined with the log-likelihood soft-decision helper data to derive the soft-decision input of the decoder algorithms as follows:

$$L_i = (-1)^{c_i'} \cdot K_i,$$

i.e. '0' code bits become positive log-likelihoods, and '1' code bits become negative ones.

Soft-Decision Decoding Algorithms

SDML repetition decoding of soft-decision information L amounts to calculating $L^* = \sum_{i=0}^{n-1} L_i$. The most likely transmitted code word was all zeros if $L^* > 0$ and all ones if $L^* < 0$. Moreover, the magnitude of L^* gives a reliability for this decision which allows us to perform a second soft-decision decoding for the outer code. Algorithm 1 outlines the simple operation for the SDML decoding of a repetition code. As an outer code, we use an $RM(r, m)$ code and decode it with an adapted version of the soft-decision GMC decoding algorithm as introduced by Schnabl and Bossert [123]. The soft-decision output of the repetition decoder is used as input by the GMC decoder. The operation of the GMC decoder we use is given by Algorithm 2. Note that this a recursive algorithm, calling itself twice if $0 < r < m$.

Decoder Architecture

We propose a resource-optimized hardware architecture to efficiently execute the soft-decision decoders given by Algorithms 1 and 2. As a general architecture, we

---

Algorithm 1 SDML-DECODE-Repetition$_n$ (L) with *soft output*

$L^* := \sum_{i=0}^{n-1} L_i$
return $(L^*, \dots, L^*)_n$

Algorithm 2 GMC-DECODE-RM$(r, m)(L)$ with *soft output*

---

    define $F(x, y) := \text{sign}(x \cdot y) \cdot \min\{\text{abs}(x), \text{abs}(y)\}$
    define $G(s, x, y) := \lfloor \frac{1}{2}(\text{sign}(s) \cdot x + y) \rfloor$
    if $r = 0$ **then**
        $L^* = $ SDML-DECODE-Repetition$_{2^m}$ (L)
    else if $r = m$ **then**
        $L^* = L$
    else
        $L_j^{(1)} = F(L_{2j-1}, L_{2j}), \forall j = 0 \dots 2^{m-1} - 1$
        $L^{(1)*} = $ GMC-DECODE-RM$(r - 1, m - 1)(L^{(1)})$
        $L_j^{(2)} = G(L_j^{(1)*}, L_{2j-1}, L_{2j}), \forall j = 0 \dots 2^{m-1} - 1$
        $L^{(2)*} = $ GMC-DECODE-RM$(r, m - 1)(L^{(2)})$
        $L^* = (F(L_0^{(1)*}, L_0^{(2)*}), L_0^{(2)*}, \dots, F(L_{2^{m-1}-1}^{(1)*}, L_{2^{m-1}-1}^{(2)*}), L_{2^{m-1}-1}^{(2)*})$
    end if
    return L^*

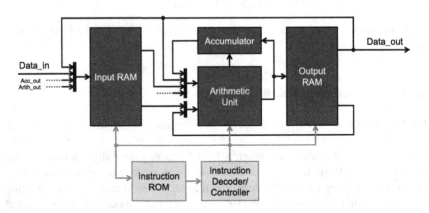

Fig. 6.2 Architecture design of the soft-decision decoder

opt for a highly serial execution of the algorithms using a small 8-bit custom datapath. The summation in Algorithm 1 is implemented by means of a serial accumulator. The functions $F(x, y)$ and $G(s, x, y)$ in Algorithm 2 are respectively evaluated in a 3-cycle execution and a 2-cycle execution. The high-level architecture is shown in Fig. 6.2. The algorithm execution is controlled by a programmable controller taking instructions from an instruction ROM, which allows support for different code parameters. The size of both data RAMs depends on the used codes.

Table 6.1 Overview of the implementation results of our reference design soft-decision secure sketch, and an efficiency comparison to an earlier proposed hard-decision secure sketch implementation

	Our design [85]	Bösch et al. [11]
FPGA slices	164	580
FPGA block RAMs	2	?
Cycles for one decoding	1248	1716
Critical path	19.9 ns	6.6 ns
SRAM PUF response size	192	264
Min-entropy after sketching	22	13

6.3.4 Implementation Results on FPGA

Table 6.1 presents the implementation results for a reference design of our soft-decision secure sketch on a Xilinx Spartan-3E500 FPGA platform. The reference design implements a concatenation of a $[3, 1, 1]$ repetition soft-decision decoder and an $RM(2, 6) \equiv [64, 22, 7]$ Reed-Muller soft-decision decoder. This combination is able to correct all errors in an SRAM PUF response with a simulated bit error rate of 15 %, with a failure rate $\leq 10^{-6}$. These parameters are chosen to objectively compare our results against the hard-decision secure sketch implementation presented by Bösch et al. [11].

From Table 6.1 it is clear that the use of soft-decision information results in a significant efficiency gain. Whereas the earlier proposed hard-decision decoder is left with 13 bits of min-entropy from a (full-entropy) 264-bit SRAM PUF response after sketching, our soft decision reference design keeps 22 bits of min-entropy from a 192-bit response. Moreover, as it turns out, the implementation of our soft-decision decoder is also significantly more area-efficient.

6.4 Practical PUF-Based Key Generation

6.4.1 Motivation

The main rationale behind what we call *practical* PUF-based key generation is that we abandon the requirement of generating information-theoretically random keys and instead aim for cryptographically secure randomness based on reasonable assumptions and best-practice techniques, aiming to gain in overall efficiency. While representing a certain sacrifice in security from a theoretical perspective, the abandonment of information-theoretical randomness can be convincingly motivated from a practical viewpoint:

- As argued in Sect. 6.2.2, the choice for information-theoretical randomness comes at a direct implementation cost for the randomness source due to the unavoidable and large min-entropy losses in the strong extractor.
- Information-theoretical randomness is conditioned on strong min-entropy guarantees for the randomness source. In practice it is very often impossible to provide a reasonably accurate estimation of a source's min-entropy, let alone a guarantee thereon. If such guarantees cannot be provided, it doesn't make much sense to consider information-theoretical randomness any further.
- For most applications, information-theoretical randomness is not strictly required. In fact, apart from a few notable exceptions, nearly all practical cryptographic primitives are themselves not information-theoretically secure operations, which makes the requirement of information-theoretical randomness a practical exaggeration.
- Most currently deployed security systems and key generators do not use information-theoretical randomness, but instead apply a more pragmatic practical approach based on seeded PRNGs, as detailed in Sect. 6.2.2.

6.4.2 Practical Key Generation from a Fuzzy Source

Practical Secure Sketch

The secure sketch, as defined in Definition 26, is already practical in the sense that it can be constructed quite efficiently, as demonstrated by the code-offset and the syndrome construction described in Sect. 6.2.1, and its efficiency is typically only limited by the efficiency of the underlying error-correcting codes. We propose a slightly more generalized definition of a secure sketch which will prove its use in the practical fuzzy extractor we describe next.

Definition 30 An $(\ell, p_e, \mu, p_{fail}, \mu')$-practical secure sketch is a pair of efficient (possibly randomized) procedures Sketch $: \{0, 1\}^\ell \to \{0, 1\}^* : y \to w$ and Recover $: \{0, 1\}^\ell \times \{0, 1\}^* \to \{0, 1\}^\ell : (y', w) \to y''$, with the following two properties:

- The correctness property of a practical secure sketch guarantees that if

$$\Pr\big(\mathbf{HD}(Y; Y') \leq t\big) \geq F_{bino}(t; \ell, p_e),$$

  then

$$\Pr\big(Y \neq \mathsf{Recover}\big(Y', \mathsf{Sketch}(Y)\big)\big) \leq p_{fail}.$$

- The security property of a practical secure sketch guarantees that if $\rho(Y) \geq \mu$ then $\rho(Y|W) \geq \mu'$.

It is clear that a practical secure sketch is very similar to a secure sketch as defined in Definition 26, and can be built with the same constructions described in

Sect. 6.2.1. However, both the correctness and the security property are slightly relaxed:

- The correctness property is no longer conditioned on a hard limit on the number
  of errors, but instead on an error rate parameter p_e which describes the probability of a number of differing bits by means of a binomial distribution. An even
  further generalization can be made by considering a generic cumulative distribution function of the errors on the fuzzy secret, but the binomial case will suffice
  for our use. Equivalently, the recovery is no longer hard-guaranteed, but is only
  certain up to a certain failure probability p_{fail}. This generalization more accurately
  captures realistic scenarios where hard bounds on error rates can only rarely be
  provided.
- The security property is also more relaxed since it is no longer conditioned on
  the min-entropy, but only on the Shannon entropy (density) of the fuzzy secret.
  Equivalently, it only provides a guarantee about the Shannon entropy (density) of
  the output.

Practical Fuzzy Extractor

In analogy to practical secure sketches, we also propose a practical fuzzy extractor:

Definition 31 An $(\ell, p_e, \mu, p_{fail}, m)$-practical fuzzy extractor is a pair of efficient
(possibly randomized) procedures

$$\text{Generate} : \{0, 1\}^{\ell} \to \{0, 1\}^m \times \{0, 1\}^* : y \to (z, w),$$

and

$$\text{Reproduce} : \{0, 1\}^{\ell} \times \{0, 1\}^* \to \{0, 1\}^m : (y', w) \to z,$$

with the following two properties:

- The correctness property of a practical key generator guarantees that if

$$\Pr\big(\mathbf{HD}(Y; Y') \le t\big) \ge F_{bino}(t; \ell, p_e),$$

and $(Z, W) \leftarrow \text{Generate}(Y)$, then

$$\Pr\big(Z \ne \text{Reproduce}(Y', W)\big) \le p_{fail}.$$

- The security property of a practical key generator states that if $\rho(Y) \ge \mu$, then
  Z is the result of an entropy accumulation over at least m bits of entropy:
  $Z \leftarrow \text{Acc}(S)$, with $H(S|W) \ge m$.

It is clear that this practical fuzzy extractor is a variant of a fuzzy extractor with
relaxed conditions. As a consequence, the practical key generator no longer produces an information-theoretically secure random output. However, through the use

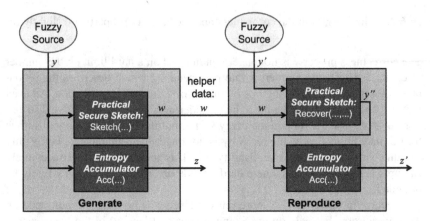

Fig. 6.3 Construction of a practical fuzzy extractor from a practical secure sketch and an entropy accumulator

of a secure entropy accumulator as described in Sect. 6.2.2, the produced output is still at least as random as that of secure cryptographic PRNGs which are currently widely used in practice to generate secure keys from low-entropy physical randomness sources.

It is evident that an $(\ell, p_e, \mu, p_{fail}, m)$-practical fuzzy extractor can be straightforwardly constructed from an $(\ell, p_e, \mu, p_{fail}, \mu')$-practical secure sketch and an m-bit entropy accumulator with $m \leq \mu'\ell$, in the same way as a regular fuzzy extractor is constructed from a secure sketch and a randomness extractor. The construction of a practical fuzzy extractor is shown in Fig. 6.3.

Design and Design Constraints of a Practical Fuzzy Extractor Construction

We propose a design of a $(\ell, p_e, \mu, p_{fail}, m)$-practical fuzzy extractor based on a syndrome construction for the practical secure sketch and an m-bit cryptographic hash function for the entropy accumulator. When the syndrome construction deploys an $[n, k, t]$ linear block code $\mathcal{C}$, a number of design constraints on the parameters of usable codes can be derived:

1. *Practicality constraint*: an efficient decoding algorithm for $\mathcal{C}[n, k, t]$ is known.
2. *Rate constraint*: $\frac{k}{n} > 1 - \mu$, and $\ell \geq r \cdot n$ with $r = \lceil \frac{m}{k - n(1-\mu)} \rceil$.
3. *Correction constraint*: $t \geq F_{bino}^{-1}((1 - p_{fail})^{\frac{1}{r}}; n, p_e)$.

Any binary linear block code which meets these constraints can be used to build the practical fuzzy extractor. It will become clear that an efficiency optimization in practice comes down to selecting an appropriate code which minimizes the required length ℓ of the input from a fuzzy source with given characteristics (p_e, μ), in order to meet the desired requirements (p_{fail}, m) for the output. This optimization hence needs to be performed over these three constraints.

The Practicality Constraint Is an evident requirement. If the used code is not efficiently decodable, the syndrome construction cannot be practically executed.

The Rate Constraint Expresses the requirement that not all unpredictability of the fuzzy secret can be leaked through the helper data. For every n bits of the fuzzy secret, a helper data string of $(n - k)$ bits is generated. The remaining conditional entropy in the n bits of the fuzzy secret is at least $n \cdot \mu - (n - k) = k - n(1 - \mu)$. To have any arguable entropy left which can be accumulated, this quantity needs to be strictly positive, from which the first part of the rate constraint immediately follows. Even if the remaining entropy per n-bit fuzzy input is positive, the construction still requires at least $r = \lceil \frac{m}{k-n(1-\mu)} \rceil$ such n-bit inputs from the fuzzy source in order to accumulate m bits of entropy.

The Correction Constraint Expresses the requirement that the used code needs to be able to correct a minimal amount of bit errors for the construction to meet the overall failure rate. The probability of a successful recovery of a single n-bit fuzzy input is at least $\Pr(\mathbf{HD}(Y^n; Y'^n) \leq t) = F_{\text{bino}}(t; n, p_e)$, and for the overall construction this is repeated r times; hence the overall success rate is at least $F_{\text{bino}}(t; n, p_e)^r \geq 1 - p_{\text{fail}}$. From this, the correction constraint on the used code immediately follows.

Extension to Concatenated Codes

Often, fuzzy sources, e.g. intrinsic PUFs, have relatively high error rates, up to 10 % and even more. Single linear block codes have difficulties in reducing such high error rates to a practically acceptable failure rate, which means that only excessively long codes and/or codes with relatively low rates can be used. This restriction can be relaxed through the use of *code concatenation*, i.e. using two (or more) block codes one after the other. Typically, a simple but robust *inner* block code brings down the relatively high error rate to a lower intermediate error probability, and a more advanced *outer* block code can then achieve a very low final failure rate much more efficiently. The power of code concatenation in the context of fuzzy extractors was first discussed by Bösch et al. [11].

We extend our proposed practical fuzzy extractor construction by deploying code concatenation. In particular, we consider using a $[n_{\text{REP}} = 2t_{\text{REP}} + 1, 1, t_{\text{REP}}]$ repetition code as the inner code. A repetition code en/decoder is trivially simple to implement with a very low overhead, and is more efficient on high error rates than more advanced block codes. The outer code can still be any $[n_2, k_2, t_2]$ binary linear block code C_2. Following the same reasoning as above, the design constraints for this extended construction then become:

1. *Practicality constraint*: an efficient decoding algorithm for $C_2[n_2, k_2, t_2]$ is known. (Efficient decoding of the repetition code is trivial.)
2. *Rate constraint*: $\frac{k_2}{n_{\text{REP}} n_2} > 1 - \mu$, and $\ell \geq r \cdot n_{\text{REP}} n_2$ with $r = \lceil \frac{m}{k_2 - n_{\text{REP}} n_2 (1 - \mu)} \rceil$.
3. *Correction constraint*: $t_2 \geq F_{\text{bino}}^{-1}((1 - p_{\text{fail}})^{\frac{1}{r}}; n, p'_e)$, with $p'_e = 1 - F_{\text{bino}}(t_{\text{REP}}; n_{\text{REP}}, p_e)$.

6.4.3 Comparison of Key Generation with Intrinsic PUFs

A practical fuzzy extractor can be directly used to generate PUF-based keys of cryptographic quality when an accurate characterization of the used PUF's response distribution in terms of binary error rate (p_e) and response entropy density (μ) is available. These are exactly the quantifiers we obtained from the eight experimentally verified intrinsic PUF constructions discussed in Chap. 4 and which are summarized in Table 4.10. We will now use these quantifiers in combination with the practical fuzzy extractor design presented in Sect. 6.4.2 to assess the key generation efficiency of the different intrinsic PUFs.

Concrete Practical Fuzzy Extractor Implementation

We consider a practical fuzzy extractor design that deploys the syndrome construction practical secure sketch based on a concatenation of a binary repetition code as an inner code and a binary BCH code as an outer code. BCH codes are a class of particularly efficient linear block codes for which efficient decoding algorithms exist. Binary BCH codes with parameters $[n_{BCH}, k_{BCH}, t_{BCH}]$ are defined for $n_{BCH} = 2^u - 1$, but through the use of code word shortening, BCH codes of any length $[n_{BCH} = 2^u - 1 - s, k_{BCH} - s, t_{BCH}]$ can be constructed. For the comparison, we assume that a generic BCH decoder is available which can efficiently decode code lengths up to $n_{BCH} = 2047$. In Sect. 6.4.4 we argue that this is a reasonable assumption by presenting a hardware design and an efficient implementation of such a decoder. For the entropy accumulator we consider an implementation of a secure cryptographic hash function which generates m-bit outputs. The output of the practical fuzzy extractor is the running hash value of the recovered fuzzy inputs.

Comparison of Optimized Key Generation Efficiencies

A PUF-based key generator is constructed from each of the eight studied intrinsic PUFs with parameters $(\hat{p}_{\mathcal{P}}^{intra}, \rho(Y^\ell))$ as summarized in Table 4.10, and a $(\ell, \hat{p}_{\mathcal{P}}^{intra}, \rho(Y^\ell), p_{fail}, m)$-practical fuzzy extractor design as presented above. For a given key requirement (p_{fail}, m), the overall construction is optimized to need the smallest possible number of PUF response bits ℓ, since the implementation cost of the PUF generally scales with ℓ. For both arbiter-based PUFs, the optimization goal is slightly different: they require the least amount of fuzzy inputs, as expressed by r in the constraints, since each input needs to be generated from a different arbiter PUF. This is a result of the decreasing entropy density of an arbiter PUF's responses when their length increases. In both cases, the optimization is performed over the constraints expressed in Sect. 6.4.2 (the extended construction with code concatenation).

The obtained optimal code selections and corresponding implementation results for each of the eight intrinsic PUF constructions, and for each of three key requirements, is presented are Table 6.2. The considered key requirements are:

1. A 128-bit generated key with failure rate at most 10^{-6};
2. A 128-bit generated key with failure rate at most 10^{-9};
3. A 256-bit generated key with failure rate at most 10^{-12}.

The required silicon area is also calculated in the same manner (and with the same disclaimers) as that described in Sect. 5.2.3. It is immediately clear that the latch PUF, the D flip-flop PUF and the basic arbiter PUF are not able to generate a key even for the most relaxed key requirements, i.e. no combination of code parameters can be found meeting the design constraints. This is either due to a too large intra-distance, a too low entropy density, or a combination of both. The SRAM PUF again performs particularly well, but so does the 2-XOR arbiter PUF. Both types of ring oscillator PUFs also manage to produce a key for all three key requirements, but at a considerably higher silicon area cost than the SRAM and the 2-XOR arbiter PUFs. The buskeeper can only produce a key for the first two requirements, and only at a very large area cost which is the result of its relatively large intra-distance.

Note that the results presented in Table 6.2 are based on the estimated entropy densities for the different PUF constructions as summarized in Table 4.10 and in Fig. 4.5, and these estimations are only upper bounds. Any more accurate (i.e. lower) entropy upper bounds will also affect these key generation results. Especially for the arbiter-based PUFs, we want to point out that the considered entropy bounds are only based on rather basic modeling attacks, and more advanced modeling attacks will result in lower entropy density estimations. The results presented in Table 6.2 should be interpreted in that context.

6.4.4 A Full-Fledged Practical Key Generator Implementation

In [87], we demonstrate the practical nature of the methods and primitives introduced in Sect. 6.4.2 by designing and implementing a complete, front-to-back PUF-based cryptographic key generation module which we have titled 'PUFKY'. A reference implementation of PUFKY is developed for a modern FPGA platform. The PUFKY module integrates:

- An implementation of an intrinsic ring oscillator PUF with Lehmer-Gray response encoding and additional entropy condensing.
- An implementation of a practical secure sketch based on the syndrome construction with concatenated linear block codes.
- An implementation of an entropy accumulator as a lightweight cryptographic hash function.
- The required communication and controller blocks which orchestrate the key generation.

Besides being a convincing proof-of-concept, the result of this development is a complete and practically usable hardware security primitive which can be immediately deployed in an FPGA-based system. Moreover, the applied modular design methodology allows for an easy reconfiguration using other PUFs, error correction

Table 6.2 PUF-based key generation results for eight intrinsic PUF constructions studied in Chap. 4. '/' means that no suitable parameter solution can be found to meet the design constraints

PUF class	n_{REP}	n_{BCH}	k_{BCH}	t_{BCH}	r	ℓ	$\log_{10} p_{fail}$	m	Silicon area (μm^2)
$m = 128$ and $p_{fail} \leq 10^{-6}$									
SRAM PUF	3	375	195	21	1	1125	−6.03	128.51	914.1
Latch PUF	/	/	/	/	/	/	/	/	/
D Flip-Flop PUF	/	/	/	/	/	/	/	/	/
Buskeeper PUF	7	2002	1045	99	2	28028	−6.82	128.04	130012.7
Arbiter PUF (basic)	/	/	/	/	/	/	/	/	/
Arbiter PUF (2-XOR)	5	490	238	30	1	2450	−6.15	128.00	2179.7
Ring Oscillator PUF (P.C.)	1	443	152	39	1	443	−6.00	128.48	52718.8
Ring Oscillator PUF (L.G.)	3	1371	678	69	1	4113	−6.01	128.09	79078.1
$m = 128$ and $p_{fail} \leq 10^{-9}$									
SRAM PUF	3	451	208	29	1	1353	−9.21	128.04	1099.4
Buskeeper PUF	9	2020	1294	73	6	109080	−9.02	128.40	505986.3
Arbiter PUF (2-XOR)	5	252	88	25	2	2520	−9.19	128.88	4359.4
Ring Oscillator PUF (P.C.)	1	510	156	51	1	510	−9.35	128.92	60250.0
Ring Oscillator PUF (L.G.)	3	1812	855	99	1	5436	−9.96	128.21	104496.1
$m = 256$ and $p_{fail} \leq 10^{-12}$									
SRAM PUF	3	853	408	49	1	2559	−12.69	256.76	2079.3
Buskeeper PUF	/	/	/	/	/	/	/	/	/
Arbiter PUF (2-XOR)	5	460	160	41	4	9200	−12.41	256.36	8718.8
Ring Oscillator PUF (P.C.)	1	994	309	89	1	994	−12.52	256.22	117675.8
Ring Oscillator PUF (L.G.)	3	1961	872	114	3	17649	−12.52	256.33	339847.7

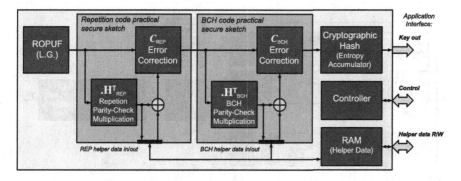

Fig. 6.4   Top-level architecture of a PUF-based key generator

methods or entropy accumulators which can be described in the model presented in this section. It is also straightforward to adapt the design parameters in order to obtain other key requirements.

Top-Level Architecture

Figure 6.4 shows the top-level architecture of the design of PUFKY, based on a concatenated repetition and BCH code. The helper data interface can be used to both read and write helper data, depending on which phase of the practical fuzzy extractor is executed (**Generate** or **Reproduce**). Only every n_{REP}th corrected bit of the repetition decoder is passed on to the BCH block, since the remaining $(n_{REP} - 1)$ bits are leaked through the repetition code's helper data and will not contribute any entropy.[1] They are discarded in the remainder of the construction.

PUF Building Block

As a PUF in the FPGA reference implementation, we take a ring oscillator PUF according to the design as described in Sect. 4.2.5. The individual oscillators are implemented in a single FPGA slice and are arranged in $b = 16$ batches of a oscillators each. The overall design of the implemented ring oscillator PUF is shown in Fig. 6.5. To generate binary response vectors, we apply the Lehmer-Gray encoding of the ordering of 16 simultaneously measured oscillator frequencies, as introduced in Sect. 4.3.1. However, we deploy two additional techniques to improve the entropy density of the generated responses: (i) frequency normalization prior to the Lehmer-Gray encoding, and (ii) entropy condensing after the encoding.

---

[1]Note that it is a convenient property of repetition codes in this situation that the entropy-bearing bits can be very straightforwardly separated from the redundant bits, which is in fact a highly efficient lossless form of randomness extraction. This is in general not possible for other codes.

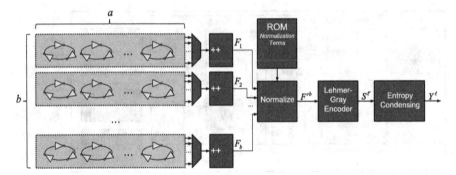

Fig. 6.5 Design of PUFKY's ring oscillator PUF with frequency normalization, Lehmer-Gray response encoding and entropy condensing

Frequency Normalization Only a portion of a measured frequency F_i will be random, and only a portion of that randomness will be caused by the effects of process variations on the considered oscillator. The analysis from Maiti et al. [91] demonstrates that F_i is subject to both device-dependent and oscillator-dependent *structural* bias. Device-dependent bias does not affect the ordering of oscillators on a single device, so we will not consider it further. Oscillator-dependent structural bias on the other hand is of concern to us since it may have a severe impact on the randomness of the frequency ordering. From a probabilistic viewpoint, it is reasonable to assume the frequencies F_i to be independent, but due to the oscillator-dependent structural bias we cannot consider them to be *identically* distributed since each F_i has a different expected value $\mathsf{E}[F_i]$. The ordering of F_i will be largely determined by the deterministic ordering of $\mathsf{E}[F_i]$ and not by the effect of random process variations on F_i. Fortunately, we are able to obtain an accurate estimate $\tilde{\mu}_{F_i}$ of $\mathsf{E}[F_i]$ by averaging over many measurements of F_i on many devices. Embedding this estimate permanently in the design and subtracting it from the measured frequency gives us a *normalized frequency* $F_i' = F_i - \tilde{\mu}_{F_i}$. Assuming $\tilde{\mu}_{F_i} \approx \mathsf{E}[F_i]$, the resulting normalized frequencies F_i' will be independent *and* identically distributed (i.i.d.).

Entropy Condensing Based on the i.i.d. assumption of the normalized frequencies, the resulting Lehmer-Gray encoding will have a full entropy of $H(S^{\ell'}) = \sum_{i=2}^{b} \log_2 i$ expressed in a vector of length $\ell' = \sum_{i=2}^{b} \lceil \log_2 i \rceil$. This means the entropy density of this vector is already quite high; for the considered design with $b = 16$, this becomes $\rho(S^{\ell'}) = \frac{44.25}{49} = 90.31$ %. However, this density can be increased further by compressing it to Y^{ℓ} with $\ell < \ell'$. Note that $S^{\ell'}$ is not quite uniform over $\{0, 1\}^{\ell'}$ since some bits of $S^{\ell'}$ are biased and/or dependent. This results from the fact that most of the Lehmer coefficients, although uniform by themselves, can take a range of values which not integer powers of two, leading to a suboptimal binary encoding. We propose a simple compression by selectively XOR-ing bits from $S^{\ell'}$ which suffer the most from bias and/or dependencies, leading to an overall increase of the entropy density. The price one pays for this *entropy condensing* operation is a slight reduction in absolute entropy of the responses, since XOR is a lossy operation, and a slight increase of the error rate of the response bits, by at most a factor $\frac{\ell'}{\ell}$. For our reference implementation, we consider a condensing to $\ell = 42$, yielding $\rho(Y^\ell) = 97.95\ \%$.

Error-Correction Building Blocks

Both the repetition code parity-check multiplication and the syndrome decoder are implemented very efficiently in a fully combinatorial manner. The parity-check multiplication of a BCH code is also implemented efficiently by means of an LFSR, which results from the special structure of a BCH code's parity-check matrix. This was discussed in Sect. 5.3.3, where we implemented such a multiplication as part of our authentication protocol. The most complex part of the practical secure sketch is the BCH syndrome decoder.

Decoding a BCH syndrome into the minimum Hamming weight error word is typically performed in three steps. First, so-called *syndrome evaluations* z_i are calculated by evaluating the syndrome as a polynomial for $\alpha, \ldots, \alpha^{2t_{BCH}}$, with α a generator for $\mathbb{F}_{2^u}$. The next step is using these z_i to generate an error location polynomial Λ. This is generally accomplished with the *Berlekamp-Massey algorithm*. First published by Berlekamp [7] and later optimized by Massey [95], this algorithm requires the inversion of an element in $\mathbb{F}_{2^u}$ in each of its $2t_{BCH}$ iterations. A less computationally intensive inversionless variant of the Berlekamp-Massey algorithm was proposed later by Burton [19]. Finally, by calculating the roots of Λ, one can find the error vector e^n. This is done with the *Chien search algorithm*, which was proposed by Chien [23], by evaluating Λ for $\alpha, \ldots, \alpha^{t_{BCH}}$. If Λ evaluates to zero for α^i then the corresponding error bit $e_{n_{BCH}-i} = 1$.

The computationally demanding BCH decoding algorithm is typically implemented with a focus on performance since it is generally applied in communication applications which require high throughput. However, in our context of a practical PUF-based key generator, an area-efficient implementation is much more desirable since only a few decodings need to be computed. In [87], we propose a BCH decoder coprocessor which is highly optimized for area efficiency, but still achieves a reasonable performance. The coprocessor is programmable with a most specialized instruction set in order to support many different code parameters with the same implementation. The presented decoder supports BCH code lengths at least up to $n_{BCH} = 2047$. The architecture of this coprocessor is shown in Fig. 6.6.

Entropy Accumulator

We use a cryptographic hash function as an entropy accumulator. For our reference implementation, we implemented the lightweight SPONGENT-128 hash function which was proposed by Bogdanov et al. [8].

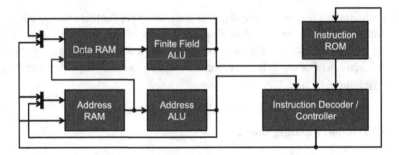

Fig. 6.6 Architecture of an area-optimized BCH decoding coprocessor

Reference Implementation

The presented PUFKY reference design was synthesized and implemented on a Xilinx Spartan-6 FPGA (XC6SLX45), which is a modern low-end FPGA in 45 nm CMOS technology, specifically targeted for embedded system solutions.

Ring Oscillator PUF Characterization   The ring oscillator PUF implementation, with $a = 64$ oscillators per batch, is characterized through an experiment on $N_{\text{puf}} = 10$ identical FPGA devices by measuring every $\ell = 42$-bit response on every device $N_{\text{meas}} = 25$ times at nominal condition. A single FPGA was also measured at two temperature corners $\alpha_{\text{L}} = (T_{env} = 10\ °\text{C})$ and $\alpha_{\text{H}} = (T_{env} = 80\ °\text{C})$ in order to assess temperature sensitivity. A conservative worst-case bit error rate estimation of 13 % (at the α_{H} corner) is obtained. The entropy density of the PUF responses is assumed to be the optimal $\rho(Y^{\ell}) = 97.95\ \%$ as a result of frequency normalization and entropy condensing.

FPGA Implementation Results   Based on the PUF characterization, we can optimize the number of required PUF response bits over the constraints for the code parameters. We aim for a key length $m = 128$ with failure rate $p_{\text{fail}} \leq 10^{-9}$. After a thorough exploration of the design space with these parameters, we converge on the following PUFKY reference implementation:

- A practical secure sketch applying a concatenation of a $[7, 1, 3]$ repetition code and a $[318, 174, 17]$ BCH code. The repetition block generates 36 bits of helper data for every 42-bit PUF response and outputs 6 bits to the BCH block. The BCH block generates 144 bits of helper data once and feeds 318 bits to the entropy accumulator. The BCH decoder is only executed once ($r = 1$). The required number of bits is $\ell = 7 \times 318 = 2226$, which can be provided by using $a = \lceil \frac{2226}{42} \rceil = 53$ oscillators per batch of the ring oscillator PUF, totalling to $53 \times 16 = 848$ oscillators for the whole ring oscillator PUF.
- The PUF generates in total $a \times \ell = 2226$ bits containing $a \times \ell \times \rho(Y^{\ell}) = 2180.4$ bits of entropy. The total helper data length is $53 \times 36 + 144 = 2052$. The remaining entropy after secure sketching is at least $2180.4 - 2052 = 128.4$ bits,

Table 6.3 Area consumption and runtime of our reference PUFKY implementation on a Xilinx Spartan-6 FPGA. Due to slice compression and glue logic the sum of module sizes is not equal to the total size. The PUF runtime is independent of clock speed

(a) Area consumption		(b) Runtimes	
Module	Size (slices)	Step of extraction	Time (cycles)
ROPUF	952	PUF output	4.59 ms
REP decoder	37	REP decoding	0
BCH syndrome calc.	72	BCH syndrome calc.	511
BCH decoder	112	BCH decoding	50320
SPONGENT-128	22	SPONGENT hashing	3990
Data RAM	38	Control overhead	489
Total	1162	*Total @ 54 MHz*	5.62 ms

which are accumulated in an $m = 128$-bit key by the SPONGENT-128 hash function implementation.

The total size of our PUFKY reference implementation for the considered FPGA platform is 1162 slices, of which 82 % is taken up by the ring oscillator PUF. Table 6.3a lists the size of each submodule used in the design. The total time spent to extract the 128-bit key is approximately 5.62 ms (at 54 MHz). Table 6.3b lists the number of cycles spent in each step of the key extraction.

6.5 Conclusion

In this chapter we have studied in detail the constructions and methodologies required to generate a cryptographically secure key in a reliable manner from a PUF response, and we have presented a variety of new techniques and insights which make PUF-based key generation more efficient and practical.

In Sect. 6.3, we have introduced an important extension to secure sketches by taking the soft-decision information of PUF response bits into account. This results in a significant improvement with regard to the entropy loss induced by secure sketches, leading to a large gain in PUF response efficiency as expressed by the results in Table 6.1. Moreover, a practical design and implementation of such a soft-decision secure sketch has been presented which turns out to be more area-efficient than all previously proposed implementations.

The notions of a *practical secure sketch* and a *practical fuzzy extractor* are introduced in Sect. 6.4. These present a number of generalizations of the traditional concepts, the most important one being the abandonment of information-theoretical security requirements in favor of the more practical and efficient technique of entropy accumulation, which is also widely used in implementations of cryptographic PRNGs. We propose an efficient generic construction of a practical secure sketch and a practical fuzzy extractor and derive a number of parameter constraints for this construction. We use these derived constraints to assess the key generation performance of the intrinsic PUFs studied in Chap. 4 in the newly proposed practical fuzzy extractor construction and present a comparative analysis in Table 6.2. From this analysis, it is clear that SRAM PUFs and also 2-XOR arbiter PUFs can be used to efficiently generate cryptographic keys.

Finally, to demonstrate the practicality of the newly proposed construction, we present a complete front-to-back PUF-based key generator design, which we have called 'PUFKY', and a reference implementation on an FPGA platform. This implementation contains a functional ring oscillator PUF with Lehmer-Gray response encoding, an efficient, practical, secure sketch implementation deploying a newly developed low-weight BCH decoder and an entropy accumulator implemented as a lightweight cryptographic hash function. This reference system generates 128-bit secure keys with a reliability $\geq 1 - 10^{-9}$, and can be directly instantiated as a hardware security IP core in an FPGA-based digital system.

Chapter 7
Conclusion and Future Work

7.1 Conclusions

PUFs are physical security primitives which enable trust in the context of digital hardware implementations of cryptographic constructions, in particular they are able to initiate physically unclonable and secure key generation and storage. In this book we have studied physically unclonable functions or PUFs, in particular: (i) their concept and constructions, (ii) their properties, and (iii) their applications as a physical root of trust, and the relations between these three.

In Chap. 2 we introduced the concept of a physically unclonable function in great detail through an extensive study and analysis of earlier work and existing constructions. Based on similar and distinguishing construction characteristics, we discussed strengths and weaknesses and possible classifications. The most noteworthy identified subclass is so-called *intrinsic PUFs*, i.e. PUF constructions which are based on internal evaluations of implicitly obtained random creation features, because they are particularly well-fit for integration in larger security systems. It is however difficult to compare the practical value of different proposals from the literature due to a wide variety of implementation technologies and experimental considerations. The lateral overview does present a good insight in recurring intrinsic PUF implementation techniques: (i) amplification of microscopic unique features through differential measurement, (ii) physical enhancement of PUF behavior through low-level (design-intensive) implementation control, and (iii) algorithmic behavioral improvement through adapted post-processing techniques.

In Chap. 3 we identified the most important usability and physical security properties attributed to PUFs and we introduced clear definitions pointing out what exactly each of these properties signify. Through a comparative analysis on a representative subset of PUF constructions and a number of important non-PUF reference cases, we discovered which properties are really *defining* for a PUF construction, and which are convenient additions but are in no way guaranteed for every PUF. As it turns out, the core defining property of a PUF is its *physical unclonability*. A second contribution of this chapter was the introduction of a formal framework

R. Maes, *Physically Unclonable Functions*, DOI 10.1007/978-3-642-41395-7_7,
© Springer-Verlag Berlin Heidelberg 2013

for using PUFs, and more general physical security primitives, in theoretical security reductions at a higher level. We defined robustness, physical unclonability and unpredictability in this framework.

With the goal of testing their security and usability characteristics as accurately and as objectively as possible, we implemented and evaluated eight different intrinsic PUF constructions on a 65 nm CMOS silicon platform. The design process and the experimental results of this ASIC chip are described in detail in Chap. 4. The measurements of each of these eight constructions were tested for their reproducibility and their uniqueness through a characterization of their respective intra- and inter-distance distributions using well-chosen statistical parameters. The unpredictability of the studied PUFs is also estimated by proposing and evaluating a number of increasingly tighter upper bounds of the entropy density of their response bitstrings. The experiments in this chapter yielded the first-ever objective comparison between multiple different intrinsic PUF constructions on the same platform, and a quantitative representation of their primary characteristics which can be directly used to assess their efficiency and performance in the following physical security applications.

The first PUF-based security applications we considered in Chap. 5 were entity identification and authentication, as defined in Sect. 5.1.1. We first elaborated on the identifiability property of PUFs and demonstrated how a fuzzy identification scheme based on PUF responses can be set up, equivalent to biometric identification. Based on the measured characteristics of the eight intrinsic PUF construction studied in Chap. 4, we evaluated their identifying capabilities based on the typical fuzzy identification performance metrics of false acceptance and rejection rates which are combined in receiver-operating characteristics and equal error rates. Next, we proposed a new mutual authentication protocol for PUF-carrying entities with significantly relaxed security conditions for the used PUF such that it can deploy any intrinsic PUF construction. The primary innovative aspect of this protocol is an atypical use of an error-correction technique which results in a secure yet very lightweight authentication solution. Based on this scheme, the authentication performance and efficiency of the eight studied intrinsic PUFs was again compared.

Next, in Chap. 6, we discussed in detail the prerequisites and security considerations for PUF-based secure key generation and storage, and presented and evaluated practical constructions and implementations. We first studied existing techniques and constructions for enhancing the reliability of fuzzy data, so-called secure sketches, and for boosting the unpredictability of a derived key, either from an information-theoretic perspective, based on strong and fuzzy extractors, or from a practical perspective, based on entropy accumulators. The first main contribution of Chap. 6 was a new and significantly improved construction of a secure sketch based on the innovative idea of using available soft-decision information of the PUF's response bits. Secondly, we introduced a practical generalization of fuzzy extractors in which we traded information-theoretical security for a large gain in efficiency, while still retaining adequate security based on widely used best-practice entropy accumulation. Again, we tested the key generation capacity of the eight studied intrinsic PUFs, based on a proposed design of a practical fuzzy extractor. Lastly, to demonstrate the feasibility and applicability of this design, we made a fully functional, efficient, yet flexible FPGA reference implementation of a PUF-based key generator, including a newly proposed ring-oscillator PUF with Lehmer-Gray response encoding, a practical secure sketch based on a new highly optimized BCH decoder, and entropy accumulation with a lightweight hash function.

7.2 Future Work

The generic concept of a physically unclonable function was presented little over a decade ago, but the major research contributions on this topic are situated in the last couple of years. PUFs are hence a relatively new subject in the field of physical security, and a number of interesting future research directions, as well as some encountered open problems can be identified.

PUF Constructions   As pointed out a number of times in this book, one of the major open questions on the side of PUF constructions is if, and how, an efficient truly unclonable intrinsic PUF, also called a strong PUF by some authors, can be built. Some candidate circuits have been proposed, but they are currently lacking any convincing argumentation of being unpredictable which is required for achieving mathematical and true unclonability. Due to the difficulty of mathematically proving unpredictability based on physical arguments, it is more convenient to assess unpredictability in relation to the best known attack. This method has been applied successfully in recent decades to cryptographic symmetric-key primitives. However, to make independent analysis possible, Kerckhoffs' principle needs to be obeyed. For physical security primitives like PUFs, this entails that the full design *and* implementation details of a proposed construction need to be disclosed. A more practical alternative is to publish, in an easily accessible electronic format, extensive datasets of experimental PUF evaluations for public scrutiny. We hope that the practical performance evaluation methods proposed in this book can contribute to more meaningful and objective assessments of the value of different PUF constructions.

Another open issue is the construction of a physically reconfigurable PUF, as discussed in Sect. 2.5.3. Rather exotic suggestions were proposed but never tested. A logically reconfigurable PUF is a good emulation from the behavioral perspective, but does not achieve the same strong physical security qualities.

Finally, as with all hardware (security) primitives, continuing research effort is required to develop increasingly more efficient constructions which offer better trade-offs and enable applications in resource-constrained environments.

PUF Properties   For PUF properties, we see the need for research effort on two levels. Firstly, on a practical level, more detailed methods and techniques need to be developed and investigated for assessing if, and to what extent, a proposed PUF construction meets a particular property. For some properties, this is based on an analysis of the inter- and intra-distance distributions of measured PUF responses, and a number of evaluation methods have been proposed in this book. Presenting detailed and accurate statistics of these distributions is hence indispensable when proposing a new PUF construction. In-depth and independent analysis of a PUF's construction and experimental results is required to assess its unpredictability, e.g. by means of tighter response entropy bounds.

For other properties, an assessment is based on physical arguments and this is again difficult to approach from a mathematical perspective. Qualities like physical unclonability and tamper evidence can, and should only be assessed in a heuristic manner. Especially the core property of physical unclonability requires special attention in that regard. A PUF which, through improved physical insights or technological progress, is found to be no longer physically unclonable, ceases to be a PUF.

On a more theoretical level, it needs to be further investigated how PUFs can be deployed as formal primitives in a security system, i.e. allow theoretical designers to reason about PUFs in a formal way without having to worry about their physical aspects, which are often beyond their expertise. We have proposed such a formal framework for describing physical functions and have applied this to PUFs and their core properties. Extending this framework to other primitives and properties will be an interesting and fruitful exercise.

PUF Applications   In this book, we have demonstrated how to achieve two primary PUF-based security objectives: entity authentication and secure key generation/storage. We see further developments possible, both in trying to achieve basic security objectives based on PUFs, as well as attempting to integrate PUFs and PUF-based primitives securely and efficiently into larger security systems. A noteworthy direction of the former which was not further discussed in this book is so-called *hardware-entangled cryptography*, i.e. designing basic cryptographic primitives which directly deploy a PUF as a building block. We introduced this approach in [2] and illustrated it with a secure design of a PUF-based block cipher.

To facilitate the further deployment of PUFs in security systems, we see the following interesting and essential future challenges:

- A greater insight and effort on the low-level silicon circuit design of intrinsic PUF constructions will result in increasingly more efficient (smaller, more robust, more random) structures.
- Both incremental and fundamental progress in the development and application of post-processing techniques required to use a PUF (reliability enhancement, randomness extraction, ...) is possible, e.g. the use of alternative error-correction techniques (non-binary codes, non-linear codes, ...) and more insight into entropy accumulation techniques.
- Other physical security qualities of PUFs need to be considered, in particular their resistance to side-channel attacks. Initial work on side-channel analysis of PUFs was recently introduced by Karakoyunlu and Sunar [62], Merli et al. [97], but more elaborate research efforts are required.

To conclude, we expect that the continuous improvements and better understanding of their practicality and of the physical as well as algorithmic security of PUFs will enable and accelerate their further transition into common and new security applications.

Appendix A
Notation and Definitions from Probability Theory and Information Theory

A.1 Probability Theory

A.1.1 Notation and Definitions

Variables and Sets In this book we use the standard notions and definitions of probability theory, such as a random event, a random variable and a probability distribution. A random variable is typically denoted as an upper case letter, Y, and a particular outcome thereof as the corresponding lower case letter, $Y = y$. The set of all outcomes of Y is denoted by a calligraphic letter, $\mathcal{Y}$. The number of elements in $\mathcal{Y}$, also called the *cardinality* of $\mathcal{Y}$, is denoted as $\#\mathcal{Y}$.

Vectors and Distances If Y is a (random) vector, we denote its length (number of elements) as $|Y|$. *Hamming distance* is a distance metric between two vectors of equal length and is defined as the number of positions in both vectors with differing values:

$$\mathbf{HD}(Y; Y') \triangleq \#\{i : Y_i \neq Y_i'\}.$$

Often, the Hamming distance is expressed as a fraction of the vectors' length, also called the *fractional Hamming distance*:

$$\mathbf{FHD}(Y; Y') \triangleq \frac{\mathbf{HD}(Y; Y')}{|Y|}.$$

The *Hamming weight* of a single vector is the number of positions with non-zero values:

$$\mathbf{HW}(Y) \triangleq \#\{i : Y_i \neq 0\}.$$

The *Euclidean distance* between two real-values vectors is defined as:

$$\|Y - Y'\| \triangleq \sum_{i=1}^{|Y|} \sqrt{\left(Y_i - Y_i'\right)^2}.$$

R. Maes, *Physically Unclonable Functions*, DOI 10.1007/978-3-642-41395-7,
© Springer-Verlag Berlin Heidelberg 2013

To make explicit that we consider a vector of length n, we also write $Y^n = (Y_1, Y_2, \ldots, Y_n)$. Sometimes, we need to refer to the first portion of a vector Y^n, up to and including the i-th element, we denote this as $Y^{(i)} = (Y_1, Y_2, \ldots, Y_i)$.

Distributions and Sampling   The probability distribution of a discrete random variable[1] is specified by its probability mass function:

$$p(y) \triangleq \Pr(Y = y),$$

or its corresponding cumulative distribution function:

$$P(y) \triangleq \Pr(Y \leq y).$$

Note that random variables, probability distributions and probability mass functions are intimately linked and are often used as synonyms of each other. We write $Y \leftarrow \mathcal{Y}$ to signify that Y is a random variable which is distributed over $\mathcal{Y}$ according to its corresponding distribution. Equivalently, we write $y \xleftarrow{\$} \mathcal{Y}$ to signify that the value y is randomly sampled from $\mathcal{Y}$ according to the distribution of Y. If no explicit distribution is given, the sampling of y is considered according to a uniform distribution over $\mathcal{Y}$.

Distribution Parameters   The expected value or *expectation* of a random variable Y is defined as:

$$E[Y] \triangleq \sum_{y \in \mathcal{Y}} y p(y),$$

and equivalently the expected value of a function of Y, $f(Y)$ is defined as:

$$E_{y \in \mathcal{Y}}[f(Y)] \triangleq \sum_{y \in \mathcal{Y}} f(y) p(y).$$

The variance of a random variable is defined as:

$$\Sigma^2[Y] \triangleq E_{y \in \mathcal{Y}}\left[(Y - E[Y])^2\right],$$

and the standard deviation of Y as:

$$\Sigma[Y] \triangleq \sqrt{\Sigma^2[Y]}.$$

Statistical Distance   The difference between the distributions of two discrete random variables A and B which take samples from the same set $\mathcal{V}$ is often quantified

---

[1]We only use discrete random variables in this work, hence we do not elaborate on continuous extensions.

by means of their *statistical distance*:

$$\mathbf{SD}(A; B) \overset{\triangle}{=} \frac{1}{2} \sum_{v \in \mathcal{V}} \left| \Pr(A = v) - \Pr(B = v) \right|.$$

Based on their statistical distance, a number of useful statistical and information-theoretic properties of the relation between two random variables can be shown. Of particular interest is the statistical distance between a random variable $V \leftarrow \mathcal{V}$ and U which is the uniformly distributed variable over $\mathcal{V}$. This is an effective measure for the *uniformity* of the distribution of V.

A.1.2 The Binomial Distribution

A discrete probability distribution which is of particular interest and recurs a number of times in this book is the binomial distribution. The binomial distribution describes the number of successes in n independent but identical Bernoulli experiments with success probability p. The probability mass function of the binomial distribution function is given by:

$$f_{\text{bino}}(t; n, p) \overset{\triangle}{=} \binom{n}{t} p^t (1 - p)^{n-t},$$

and the corresponding cumulative distribution function by:

$$F_{\text{bino}}(t; n, p) \overset{\triangle}{=} \sum_{i=0}^{t} f_{\text{bino}}(t; n, p).$$

The expectation of a binomially distributed random variable Y is given by $\mathsf{E}[Y] = np$ and its variance by $\Sigma^2[Y] = np(1 - p)$.

A.2 Information Theory

A.2.1 Basics of Information Theory

In this book, we use the standard information-theoretical entropy and mutual information definitions and relations, e.g. as given by Cover and Thomas [27].

Entropy is a measure of the uncertainty one has about the outcome of a random variable distributed according to a given distribution. In that sense, it quantifies the average amount of *information* one learns when observing the outcome. In the context of security, entropy is also used to express the unpredictability of a secret random variable. The concept of entropy as a measure of information was introduced by Shannon [127] and is therefore also called *Shannon entropy*.

Definition 32 The (Shannon) entropy $H(Y)$ of a discrete random variable $Y \leftarrow \mathcal{Y}$ is defined as:

$$H(Y) \overset{\triangle}{=} - \sum_{y \in \mathcal{Y}} p(y) \cdot \log_2 p(y).$$

With, by convention, $0 \cdot \log_2 0 \equiv 0$.

The entropy of a binary random variable $Y \leftarrow \{0, 1\}$, with $p(y = 1) = p$ and $p(y = 0) = (1 - p)$, becomes $H(Y) = -p \log_2 p - (1 - p) \log_2 (1 - p)$. We call this the *binary entropy function* $h(p)$:

$$h(p) \overset{\triangle}{=} -p \log_2 p - (1 - p) \log_2 (1 - p).$$

For random vectors Y^n it is sometimes convenient to express the average entropy per element, or its *entropy density*. We denote this as

$$\rho\left(Y^n\right) \overset{\triangle}{=} \frac{H(Y^n)}{n}.$$

Definition 33 The joint entropy $H(Y_1, Y_2)$ of two discrete random variables $Y_1 \leftarrow \mathcal{Y}_1$ and $Y_2 \leftarrow \mathcal{Y}_2$ with a joint distribution given by $p(y_1, y_2)$ is defined as:

$$H(Y_1, Y_2) \overset{\triangle}{=} - \sum_{y_1 \in \mathcal{Y}_1} \sum_{y_2 \in \mathcal{Y}_2} p(y_1, y_2) \cdot \log_2 p(y_1, y_2).$$

Definition 34 The conditional entropy $H(Y_2 | Y_1)$ of Y_2 given Y_1 is defined as:

$$H(Y_2 | Y_1) \overset{\triangle}{=} \mathsf{E}_{y_1 \in \mathcal{Y}_1} \left[H(Y_2 | Y_1 = y_1) \right]$$

$$= - \sum_{y_1 \in \mathcal{Y}_1} \sum_{y_2 \in \mathcal{Y}_2} p(y_1, y_2) \cdot \log_2 p(y_2 | y_1).$$

Theorem 1 *The chain rule for entropy states that:*

$$H(Y_1, Y_2) = H(Y_1) + H(Y_2 | Y_1),$$

and more generally for an arbitrary collection of random variables:

$$H\left(Y^n\right) = H(Y_1, Y_2, \dots, Y_n) = \sum_{i=1}^{n} H(Y_i | Y_{i-1}, \dots, Y_1).$$

Definition 35 The mutual information $I(Y_1; Y_2)$ between two random variables Y_1 and Y_2 is defined as:

$$I(Y_1; Y_2) \overset{\triangle}{=} \sum_{y_1 \in \mathcal{Y}_1} \sum_{y_2 \in \mathcal{Y}_2} p(y_1, y_2) \cdot \log_2 \frac{p(y_1, y_2)}{p(y_1) p(y_2)}.$$

The mutual information between two random variables is a measure for the amount of information which is shared by both variables, or in other words, the average amount of information (reduction in uncertainty) we learn about one variable when observing the other, and vice versa.

Theorem 2 *The following relations exist between entropy and mutual information*:

$$I(Y_1; Y_2) = I(Y_2; Y_1),$$

$$I(Y_1; Y_1) = H(Y_1),$$

$$I(Y_1; Y_2) = H(Y_1) - H(Y_1|Y_2) = H(Y_2) - H(Y_2|Y_1),$$

$$I(Y_1; Y_2) = H(Y_1) + H(Y_2) - H(Y_1, Y_2).$$

Definition 36 The conditional mutual information $I(Y_1; Y_2|Y_3)$ between Y_1 and Y_2 given a third random variable Y_3 is defined as:

$$I(Y_1; Y_2|Y_3) \overset{\triangle}{=} H(Y_1|Y_3) - H(Y_1|Y_2, Y_3).$$

Theorem 3 *The chain rule for mutual information states that*:

$$I(Y^n; Y_{n+1}) = I(Y_1, \ldots, Y_n; Y_{n+1}) = \sum_{i=1}^{n} I(Y_i; Y_{n+1}|Y_{i-1}, \ldots, Y_1).$$

A.2.2   Min-entropy

Min-entropy, as introduced by Rényi [109], is a more pessimistic notion of the uncertainty of a random variable than Shannon entropy. We use the definitions of min-entropy and *average* min-entropy as defined by Dodis et al. [33].

Definition 37 The min-entropy $H_\infty(Y)$ of a random variable $Y \leftarrow \mathcal{Y}$ is defined as:

$$H_\infty(Y) \overset{\triangle}{=} - \log_2 \max_{y \in \mathcal{Y}} p(y).$$

If Y is uniformly distributed, its min-entropy is equal to its Shannon entropy: $H_\infty(Y) = H(Y) = \log_2 \#\mathcal{Y}$. For other distributions, min-entropy is strictly upper-bounded by Shannon entropy: $H_\infty(Y) < H(Y)$. For two independent random variables Y_1 and Y_2 it holds that $H_\infty(Y_1, Y_2) = H_\infty(Y_1) + H_\infty(Y_2)$. However, when Y_1 and Y_2 are not independent, no prior statement can be made about this relation: $H_\infty(Y_1, Y_2) \lesseqgtr H_\infty(Y_1) + H_\infty(Y_2)$.

Definition 38 The average min-entropy $\widetilde{H}_\infty(Y|W)$ of Y given W is defined as:

$$\widetilde{H}_\infty(Y|W) \overset{\triangle}{=} - \log_2 \mathsf{E}_{w \in \mathcal{W}} \left[2^{-H_\infty(Y|W=w)} \right].$$

Note that this definition of *average* min-entropy differs from the way *conditional* entropy is typically defined in Definition 34, in particular the order of the expectation operator and the logarithm operator are reversed. To make this distinction explicit, the tilde sign ($\tilde{\ }$) is added to the average min-entropy operator: $\tilde{H}_\infty(\cdot)$. The following useful lemma about average min-entropy is proven by Dodis et al. [33]:

Lemma 3 *If $W \leftarrow \mathcal{W}$ and $\#\mathcal{W} \leq 2^\lambda$, then:*

$$\tilde{H}_\infty(Y|W) \geq H_\infty(Y, W) - \lambda \geq H_\infty(Y) - \lambda.$$

Appendix B
Non-intrinsic PUF(-like) Constructions

In this appendix we discuss *non-intrinsic* PUF constructions proposed in the literature. According to the classification proposed in Sect. 2.3.3, they are labelled 'non-intrinsic' because they either are not completely integrated in an embedding device, or they are not produced in the standard manufacturing process of their embedding device, or both. A great variety of different constructions which could be listed here have been proposed over time. Some even long before the concept of a PUF was introduced. We made an extensive selection and have grouped them based on their operating principles. Section B.1 lists a number of PUF proposals which use optical effects to obtain PUF behavior. Section B.2 describes PUFs using radio-frequency effects and Sect. B.3 lists non-intrinsic PUFs based on electronics. In Sect. B.4, we discuss a number of unrelated proposals which are of independent interest.

B.1 Optics-Based PUFs

B.1.1 Optical PUF

The interaction of visible light with a randomized microstructure quickly becomes very complex to describe, as it is a result of several physical phenomena such as absorption, transmission, reflection and scattering. If the microstructure contains random elements, the resulting interaction will often also contain a high degree of randomness and unpredictability. Optical PUFs capture the result of such a complex interaction and use it as a PUF response.

Well before the introduction of the PUF concept, Simmons [131], Tolk [142] proposed an unclonable identification system based on random optical reflection patterns. These so-called reflective particle tags were specifically developed for the identification of strategic arms in arms control treaties. A similar solution for smart-cards based on an optical token with a unique three-dimensional reflection pattern

R. Maes, *Physically Unclonable Functions*, DOI 10.1007/978-3-642-41395-7,
© Springer-Verlag Berlin Heidelberg 2013

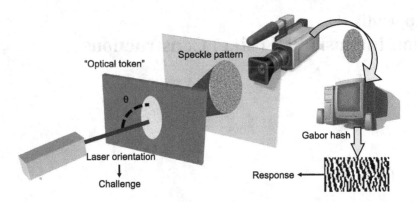

Fig. B.1 Operation of the optical PUF as proposed by Pappu et al. [105]

and stereoscopic measurement was commercialized already in 1994 by Unicate BV as 3DAS® [148].

An optical PUF based on a transparent medium was proposed by Pappu et al. [104, 105] as a *physical one-way function (POWF)*. They construct tokens containing an optical microstructure consisting of microscopic refractive particles randomly mixed in a small transparent epoxy plate. When irradiated with a laser, the token produces an irregular random speckle pattern due to the multiple scattering of the beam with the refractive particles. The pattern is digitally captured and processed into a binary feature vector using an image processing technique known as Gabor hashing. The resulting feature vector turns out to be highly unique for every token and is moreover very sensitive to minute changes in the relative positioning of the token and the laser beam. The authors propose to use different orientations of the laser to produce multiple vectors per token. In PUF terminology, the laser positioning is considered the PUF challenge and the resulting feature vector the response. The operation of this optical PUF is shown in Fig. B.1.

In his thesis, Pappu [104, Chap. 3] lists a number of earlier proposed similar constructions which served as inspiration. Following the introduction of the optical PUF by Pappu et al. [105], the construction and security aspects of optical PUFs were further studied by Ignatenko et al. [56], Škorić et al. [133], Tuyls et al. [146]. A more practical integrated design of an optical PUF was proposed by Gassend [40, Sect. 3.1.3] and also by Tuyls and Škorić [145], but no known implementations exist.

It is clear that the use of an optical PUF as proposed by Pappu et al. [105] is rather laborious, requiring a large and expensive external measurement setup involving a laser and a tedious mechanical positioning system. The reliability of optical PUF responses is also relatively low in comparison to other proposals. However, in many ways this optical PUF construction can be considered as the *prototype PUF*. The physical functionality of the PUF is obvious, including a well-defined physical challenge and response, and all desirable PUF properties are achieved: tokens produced by the same process exhibit responses with a high level of uniqueness, yet responses can be relatively easily and reliably measured. These optical PUFs also have a very large challenge space, and it was shown [133, 146] that predicting unseen responses for a particular token is computationally hard, even when an adversary has a lot of information about the token. Pappu et al. [105] moreover claim that their PUF exhibits a degree of one-wayness, hence their name physical one-way function, however this is not one-wayness in the cryptographic sense. Ultimately, it was shown [105] that the tokens are *tamper evident*, i.e. a physical modification of the token, e.g. drilling a microscopic hole in it, alters its expected responses by nearly 50 %.

From the discussion in Chap. 3, it will be clear that no single other PUF construction achieves all these properties at once. In this respect, the optical PUF can be considered a benchmark for PUF properties. The goal of all following PUF proposals is to match the properties of an optical PUF as closely as possible, but using a more integrated and practical construction.

B.1.2 Paper-Based PUFs

The basic idea behind all paper-based PUFs is scanning the unique and random fiber microstructure of regular or modified paper. As with the optical PUF, also for paper PUFs there were a number of early proposals, among others by Bauder [5], the Commission on Engineering and Technical Systems (CETS) [25], and Simmons [131], well before the introduction of the PUF concept. These were mainly considered as an anti-counterfeiting measure for currency notes. Buchanan et al. [16] propose a construction where the reflection of a focused laser beam of the irregular fiber structure of a paper document is used as a fingerprint of that document to prevent forgery. A similar approach is used by Bulens et al. [17], but they explicitly introduce ultraviolet fibers in the paper during the manufacturing process. This makes it possible to measure the paper structure using a regular desktop scanner instead of an expensive laser. Bulens et al. [17] also introduce a method to strongly link the data on the document with the paper by constructing a combined digital signature of the data and the paper's fingerprint, which is printed on the document. A very similar concept was already proposed much earlier by Simmons [131] based on quadratic residues. Recently, Sharma et al. [128] proposed a technique which enables to read unique microfeatures of regular paper using low-cost commodity equipment. They also give a good overview of many earlier proposed similar constructions. Yamakoshi et al. [155] describe a very similar concept using infrared scans of regular paper.

B.1.3 Phosphor PUF

Chong et al. [24, 59] propose to randomly blend small phosphor particles in the material of a product or its cover. The resulting random phosphorescent pattern can be detected using regular optical equipment and used as a unique fingerprint to identify the product.

B.2 RF-Based PUFs

B.2.1 RF-DNA

A construction called RF-DNA was proposed by Dejean and Kirovski [29]. They construct a small inexpensive token comparable to that used for the optical PUF, but now consisting of a flexible silicon sealant containing thin randomly arranged copper wires. Instead of observing the scattering of light as with optical PUFs, the near-field scattering of electromagnetic waves by the copper wires at other wavelengths is observed, notably in the 5∼6 GHz band. The random scattering effects are measured by a prototype scanner consisting of a matrix of RF antennas.

B.2.2 LC PUF

An LC PUF as proposed by Guajardo et al. [46] is a small glass plate with a metal plate on each side, forming a capacitor, serially chained with a metal coil on the plate acting as an inductive component. Together they form a passive LC resonator circuit which will absorb an amount of power when placed in an RF field. A frequency sweep reveals the resonance frequencies of the circuit, which depend on the exact values of the capacitive and inductive component. Due to manufacturing variations in the construction of the tokens, this resonance peak will be slightly different and unique for equally constructed LC PUFs.

B.3 Electronics-Based PUFs

B.3.1 Coating PUF

Coating PUFs were introduced by Tuyls et al. [147] and consider the randomness of capacitance measurements in comb-shaped sensors in the top metal layer of an integrated circuit. The construction of this coating PUF is shown in Fig. B.2. Instead of relying solely on the random effects of manufacturing variability, random elements are explicitly introduced by means of a passive dielectric coating sprayed directly on top of the sensors. Moreover, since this coating is opaque and chemically inert, it offers strong protection against physical attacks as well. An experimental security evaluation [147] reveals that the coating PUF is also tamper evident, i.e. after an attack with a focused ion beam (FIB) the responses of the PUF are significantly changed. A more theoretical evaluation of coating PUFs was done by Škorić et al. [134].

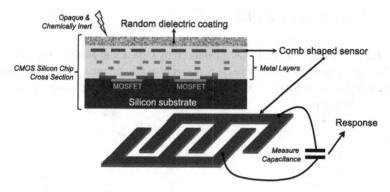

Fig. B.2 Construction of a coating PUF as proposed by Tuyls et al. [147]

B.3.2 Power Distribution Network PUF

Helinski et al. [50] propose a PUF based on the resistance variations in the power grid of a silicon chip. Voltage drops and equivalent resistances in the power distribution system are measured using external instruments and it is observed that these electrical parameters are affected by random manufacturing variability.

B.4 More Non-intrinsic PUFs

B.4.1 CD-Based PUF

Hammouri et al. [49] observed that the measured lengths of lands and pits on a regular compact disk (CD) exhibit a random deviation from their intended lengths due to probabilistic variations in the manufacturing process. Moreover, this deviation is even large enough to be observed by monitoring the electrical signal of the photodetector in a regular CD player. This was tested for a large number of CDs and locations on every CD. After an elaborate quantization procedure, an average intra-distance of $\mu_{\mathcal{P}}^{\text{intra}} = 8\,\%$ and an average inter-distance of $\mu_{\mathcal{P}}^{\text{inter}} = 54\,\%$ on the obtained bit strings is achieved.

B.4.2 Acoustical PUF

Acoustical delay lines are components used to delay electrical signals. They convert an alternating electrical signal into a mechanical vibration and back. Acoustical PUFs, as proposed by Vrijaldenhoven [154], are constructed by observing the characteristic frequency spectrum of an acoustical delay line. A bit string is extracted by performing principal component analysis (PCA).

B.4.3 Magstripe-Based PUF

The concept of a magnetic PUF was introduced by Indeck and Muller [57]. They use the inherent uniqueness of the particle patterns in magnetic media, e.g. in magnetic swipe cards. Magstripe-based PUFs are used in a commercial application to prevent credit card fraud [88].

References

1. Anderson, J. (2010). A PUF design for secure FPGA-based embedded systems. In *Asia and South-Pacific design automation conference—ASP-DAC 2010* (pp. 1–6). New York: IEEE.
2. Armknecht, F., Maes, R., Sadeghi, A.-R., Sunar, B., & Tuyls, P. (2009). Memory leakage-resilient encryption based on physically unclonable functions. In *Lecture notes in computer science (LNCS): Vol. 5912. Advances in cryptology—ASIACRYPT 2009* (pp. 685–702). Berlin: Springer.
3. Armknecht, F., Maes, R., Sadeghi, A.-R., Standaert, F.-X., & Wachsmann, C. (2011). A formal foundation for the security features of physical functions. In *IEEE symposium on security and privacy—SP 2011* (pp. 397–412). New York: IEEE.
4. Barker, E., & Kelsey, J. (2012). Recommendation for random number generation using deterministic random bit generators. NIST special publication 800-90A. http://csrc.nist.gov/publications/nistpubs/800-90A/SP800-90A.pdf.
5. Bauder, D. (1983). *An anti-counterfeiting concept for currency systems* (Technical Report PTK-11990). Albuquerque, NM, USA: Sandia National Labs.
6. Beckmann, N., & Potkonjak, M. (2009). Hardware-based public-key cryptography with public physically unclonable functions. In *Lecture notes in computer science (LNCS): Vol. 5806. International workshop on information hiding—IH 2009* (pp. 206–220). Berlin: Springer.
7. Berlekamp, E. (1965). On decoding binary Bose-Chadhuri-Hocquenghem codes. *IEEE Transactions on Information Theory, 11*(4), 577–579.
8. Bogdanov, A., Knežević, M., Leander, G., Toz, D., Varici, K., & Verbauwhede, I. (2011). SPONGENT: a lightweight hash function. In *Lecture notes in computer science (LNCS): Vol. 6917. Workshop on cryptographic hardware and embedded systems—CHES 2011* (pp. 312–325). Berlin: Springer.
9. Bolotnyy, L., & Robins, G. (2007). Physically unclonable function-based security and privacy in RFID systems. In *IEEE international conference on pervasive computing and communications—PERCOM 2007* (pp. 211–220). New York: IEEE.
10. Boneh, D., DeMillo, R. A., & Lipton, R. J. (1997). On the importance of checking cryptographic protocols for faults. In *Lecture notes in computer science (LNCS): Vol. 1233. Advances in cryptology—EUROCRYPT 1997* (pp. 37–51). Berlin: Springer.
11. Bösch, C., Guajardo, J., Sadeghi, A.-R., Shokrollahi, J., & Tuyls, P. (2008). Efficient helper data key extractor on FPGAs. In *Lecture notes in computer science (LNCS): Vol. 5154. Workshop on cryptographic hardware and embedded systems—CHES 2008* (pp. 181–197). Berlin: Springer.
12. Bose, R. C., & Ray-Chaudhuri, D. K. (1960). On a class of error correcting binary group codes. *Information and Control, 3*(1), 68–79.
13. Boyen, X. (2004). Reusable cryptographic fuzzy extractors. In *ACM conference on computer and communications security—CCS 2004* (pp. 82–91). New York: ACM.

R. Maes, *Physically Unclonable Functions*, DOI 10.1007/978-3-642-41395-7,
© Springer-Verlag Berlin Heidelberg 2013

14. Bringer, J., Chabanne, H., & Icart, T. (2009). On physical obfuscation of cryptographic algorithms. In *Lecture notes in computer science (LNCS): Vol. 5922. International conference on cryptology in India—INDOCRYPT 2009* (pp. 88–103). Berlin: Springer.

15. Brzuska, C., Fischlin, M., Schröder, H., & Katzenbeisser, S. (2011). Physically uncloneable functions in the universal composition framework. In *Lecture notes in computer science (LNCS): Vol. 6841. Advances in cryptology—CRYPTO 2011* (pp. 51–70). Berlin: Springer.

16. Buchanan, J. D. R., Cowburn, R. P., Jausovec, A.-V., Petit, D., Seem, P., Xiong, G., Atkinson, D., Fenton, K., Allwood, D. A., & Bryan, M. T. (2005). Forgery: 'Fingerprinting' documents and packaging. *Nature, 436*(7050), 475.

17. Bulens, P., Standaert, F.-X., & Quisquater, J.-J. (2010). How to strongly link data and its medium: the paper case. *IET Information Security, 4*(3), 125–136.

18. Burr, A. (2001). *Modulation and coding for wireless communications*. Upper Saddle River: Pearson.

19. Burton, H. (1971). Inversionless decoding of binary BCH codes. *IEEE Transactions on Information Theory, 17*(4), 464–466.

20. Canetti, R. (2001). Universally composable security: a new paradigm for cryptographic protocols. In *IEEE symposium on foundations of computer science—FOCS 2001* (pp. 136–145). New York: IEEE.

21. Carter, J. L., & Wegman, M. N. (1977). Universal classes of hash functions. In *ACM symposium on theory of computing—STOC 1977* (pp. 106–112). New York: ACM.

22. Chen, Q., Csaba, G., Lugli, P., Schlichtmann, U., & Ruhrmair, U. (2011). The bistable ring PUF: a new architecture for strong physical unclonable functions. In *IEEE international symposium on hardware-oriented security and trust—HOST 2011* (pp. 134–141). New York: IEEE.

23. Chien, R. (1964). Cyclic decoding procedures for Bose-Chaudhuri-Hocquenghem codes. *IEEE Transactions on Information Theory, 10*(4), 357–363.

24. Chong, C. N., Jiang, D., Zhang, J., & Guo, L. (2008). Anti-counterfeiting with a random pattern. In *International conference on emerging security information, systems and technologies—SECURWARE 2008* (pp. 146–153). New York: IEEE.

25. Commission on Engineering and Technical Systems (CETS) (1993). *Counterfeit deterrent features for the next-generation currency design* (Appendix E). Washington: National Academic Press.

26. Cortes, C., & Vapnik, V. (1995). Support-vector networks. *Machine Learning, 20*(3), 273–297.

27. Cover, T. M., & Thomas, J. A. (2006). *Elements of information theory*. New York: Wiley.

28. Daemen, J., & Rijmen, V. (2002). *The design of Rijndael*. Berlin: Springer.

29. Dejean, G., & Kirovski, D. (2007). RF-DNA: radio-frequency certificates of authenticity. In *Lecture notes in computer science (LNCS): Vol. 4727. Workshop on cryptographic hardware and embedded systems—CHES 2007* (pp. 346–363). Berlin: Springer.

30. Devadas, S., Suh, E., Paral, S., Sowell, R., Ziola, T., & Khandelwal, V. (2008). Design and implementation of PUF-based "unclonable" RFID ICs for anti-counterfeiting and security applications. In *IEEE international conference on RFID—RFID 2008* (pp. 58–64). New York: IEEE.

31. Diffie, W., & Hellman, M. E. (1976). New directions in cryptography. *IEEE Transactions on Information Theory, 22*(6), 644–654.

32. Dodis, Y., Reyzin, L., & Smith, A. (2004). Fuzzy extractors: how to generate strong keys from biometrics and other noisy data. In *Lecture notes in computer science (LNCS): Vol. 3027. Advances in cryptology—EUROCRYPT 2004* (pp. 523–540). Berlin: Springer.

33. Dodis, Y., Ostrovsky, R., Reyzin, L., & Smith, A. (2008). Fuzzy extractors: how to generate strong keys from biometrics and other noisy data. *SIAM Journal on Computing, 38*(1), 97–139.

34. Eastlake, D., Schiller, J., & Crocker, S. (2005). Randomness requirements for security. IETF RFC 4086. http://www.ietf.org/rfc/rfc4086.txt.

35. English Wiktionary: -able (2012). http://en.wiktionary.org/wiki/-able.

36. Ferguson, N., & Schneier, B. (2003). *Practical cryptography*. New York: Wiley.
37. Fischer, V., & Drutarovský, M. (2002). True random number generator embedded in reconfigurable hardware. In *Lecture notes in computer science (LNCS): Vol. 2523. Workshop on cryptographic hardware and embedded systems—CHES 2002* (pp. 415–430). Berlin: Springer.
38. Fujiwara, H., Yabuuchi, M., Nakano, H., Kawai, H., Nii, K., & Arimoto, K. (2011). A chip-ID generating circuit for dependable LSI using random address errors on embedded SRAM and on-chip memory BIST. In *Symposium on VLSI circuits—VLSIC 2011* (pp. 76–77). New York: IEEE.
39. Gallager, R. G. (1962). Low density parity-check codes. *IRE Transactions on Information Theory, 8,* 21–28.
40. Gassend, B. (2003). *Physical random functions*. M.S. Thesis, Massachusetts Institute of Technology (MIT), MA, USA.
41. Gassend, B., Clarke, D., van Dijk, M., & Devadas, S. (2002). Controlled physical random functions. In *Annual computer security applications conference—ACSAC 2002*. New York: IEEE.
42. Gassend, B., Clarke, D., van Dijk, M., & Devadas, S. (2002). Silicon physical random functions. In *ACM conference on computer and communications security—CCS 2002* (pp. 148–160). New York: ACM.
43. Gassend, B., Lim, D., Clarke, D., van Dijk, M., & Devadas, S. (2004). Identification and authentication of integrated circuits: research articles. *Concurrency and Computation: Practice and Experience, 16*(11), 1077–1098.
44. Gray, F. (1947). *Pulse code communication*. US Patent No. 2,632,058.
45. Guajardo, J., Kumar, S. S., Schrijen, G. J., & Tuyls, P. (2007). FPGA intrinsic PUFs and their use for IP protection. In *Lecture notes in computer science (LNCS): Vol. 4727. Workshop on cryptographic hardware and embedded systems—CHES 2007* (pp. 63–80). Berlin: Springer.
46. Guajardo, J., Škorić, B., Tuyls, P., Kumar, S. S., Bel, T., Blom, A. H., & Schrijen, G.-J. (2009). Anti-counterfeiting, key distribution, and key storage in an ambient world via physical unclonable functions. *Information Systems Frontiers, 11*(1), 19–41.
47. Gutmann, P. (2004). *Cryptographic security architecture*. Berlin: Springer.
48. Hammouri, G., Öztürk, E., Birand, B., & Sunar, B. (2008). Unclonable lightweight authentication scheme. In *International conference on information, communications, and signal processing—ICICS 2008* (pp. 33–48). New York: IEEE.
49. Hammouri, G., Dana, A., & Sunar, B. (2009). CDs have fingerprints too. In *Lecture notes in computer science (LNCS): Vol. 5747. Workshop on cryptographic hardware and embedded systems—CHES 2009* (pp. 348–362). Berlin: Springer.
50. Helinski, R., Acharyya, D., & Plusquellic, J. (2009). A physical unclonable function defined using power distribution system equivalent resistance variations. In *Design automation conference—DAC 2009* (pp. 676–681). New York: ACM.
51. Hocquenghem, A. (1960). Codes correcteurs d'Erreurs. *Chiffres, 2,* 147–156.
52. Holcomb, D. E., Burleson, W. P., & Fu, K. (2007). Initial SRAM state as a fingerprint and source of true random numbers for RFID tags. In *Workshop on RFID security and privacy—RFIDSec 2007*. New York: IEEE.
53. Holcomb, D. E., Burleson, W. P., & Fu, K. (2009). Power-up SRAM state as an identifying fingerprint and source of true random numbers. *IEEE Transactions on Computers, 58*(9), 1198–1210.
54. Hopper, N., & Blum, M. (2000). *A secure human-computer authentication scheme* (Technical Report CMU-CS-00-139). Pittsburgh, PA, USA: School of Computer Science, Carnegie Mellon University.
55. Hospodar, G., Maes, R., & Verbauwhede, I. (2012). Machine learning attacks on 65 nm arbiter PUFs: accurate modeling poses strict bounds on usability. In *IEEE international workshop on information forensics and security—WIFS 2012* (pp. 37–42). New York: IEEE.
56. Ignatenko, T., Schrijen, G.-J., Škorić, B., Tuyls, P., & Willems, F. M. J. (2006). Estimating the secrecy rate of physical unclonable functions with the context-tree weighting method.

In *IEEE international symposium on information theory—ISIT 2006* (pp. 499–503). New York: IEEE.

57. Indeck, R. S., & Muller, M. W. (1994). *Method and apparatus for fingerprinting magnetic media*. US Patent No. 5,365,586.

58. Java™ Platform Standard Ed. 6: Interface Cloneable (2012). http://docs.oracle.com/javase/6/docs/api/java/lang/Cloneable.html.

59. Jiang, D., & Chong, C. N. (2008). Anti-counterfeiting using phosphor PUF. In *International conference on anti-counterfeiting, security and identification—ASID 2008* (pp. 59–62). New York: IEEE.

60. Juels, A., & Sudan, M. (2006). A fuzzy vault scheme. *Designs, Codes and Cryptography, 38*(2), 237–257.

61. Juels, A., & Wattenberg, M. (1999). A fuzzy commitment scheme. In *ACM conference on computer and communications security—CCS 1999* (pp. 28–36). New York: ACM.

62. Karakoyunlu, D., & Sunar, B. (2010). Differential template attacks on PUF enabled cryptographic devices. In *IEEE international workshop on information forensics and security—WIFS 2010* (pp. 1–6). New York: IEEE.

63. Kardas, S., Akgun, M., Kiraz, M. S., & Demirci, H. (2011). Cryptanalysis of lightweight mutual authentication and ownership transfer for RFID systems. In *Workshop on lightweight security and privacy: devices, protocols, and applications—LightSec 2011* (pp. 20–25). New York: IEEE.

64. Katzenbeisser, S., Koçabas, U., van der Leest, V., Sadeghi, A.-R., Schrijen, G.-J., Schröder, H., & Wachsmann, C. (2011). Recyclable PUFs: logically reconfigurable PUFs. In *Lecture notes in computer science (LNCS): Vol. 6917. Workshop on cryptographic hardware and embedded systems—CHES 2011* (pp. 374–389). Berlin: Springer.

65. Kelsey, J., Schneier, B., & Ferguson, N. (1999). Yarrow-160: notes on the design and analysis of the Yarrow cryptographic pseudorandom number generator. In *Lecture notes in computer science (LNCS): Vol. 1758. International workshop on selected areas in cryptography—SAC 1999* (pp. 13–33). Berlin: Springer.

66. Kerckhoffs, A. (1883). La cryptographie militaire. *Journal des Sciences Militaires, IX*, 5–83.

67. Kim, I., Maiti, A., Nazhandali, L., Schaumont, P., Vivekraja, V., & Zhang, H. (2010). From statistics to circuits: foundations for future physical unclonable functions. In A.-R. Sadeghi & D. Naccache (Eds.), *Information security and cryptography. Towards hardware-intrinsic security* (pp. 55–78). Berlin: Springer.

68. Kocher, P. C. (1996). *Timing attacks on implementations of Diffie-Hellman, RSA, DSS, and other systems*. In *Lecture notes in computer science (LNCS): Vol. 1109. Advances in Cryptology—CRYPTO 1996* (pp. 104–113). Berlin: Springer.

69. Kocher, P. C., Jaffe, J., & Jun, B. (1999). Differential power analysis. In *Lecture notes in computer science (LNCS): Vol. 1666. Advances in cryptology—CRYPTO 1999* (pp. 388–397). Berlin: Springer.

70. Krishna, A., Narasimhan, S., Wang, X., & Bhunia, S. (2011). MECCA: a robust low-overhead PUF using embedded memory array. In *Lecture notes in computer science (LNCS): Vol. 6917. Workshop on cryptographic hardware and embedded systems—CHES 2011* (pp. 407–420). Berlin: Springer.

71. Kulseng, L., Yu, Z., Wei, Y., & Guan, Y. (2010). Lightweight mutual authentication and ownership transfer for RFID systems. In *IEEE international conference on computer communications—INFOCOM 2010* (pp. 1–5). New York: IEEE.

72. Kumar, S., Guajardo, J., Maes, R., Schrijen, G.-J., & Tuyls, P. (2008). Extended abstract: the butterfly PUF protecting IP on every FPGA. In *IEEE international symposium on hardware-oriented security and trust—HOST 2008* (pp. 67–70). New York: IEEE.

73. Kursawe, K., Sadeghi, A.-R., Schellekens, D., Tuyls, P., & Škorić, B. (2009). Reconfigurable physical unclonable functions—enabling technology for tamper-resistant storage. In *IEEE international symposium on hardware-oriented security and trust—HOST 2009* (pp. 22–29). New York: IEEE.

74. Lao, Y., & Parhi, K. (2011). Reconfigurable architectures for silicon physical unclonable functions. In *IEEE international conference on electro/information technology—EIT 2011* (pp. 1–7). New York: IEEE.
75. Lee, J. W., Lim, D., Gassend, B., Suh, G. E., van Dijk, M., & Devadas, S. (2004). A technique to build a secret key in integrated circuits for identification and authentication application. In *Symposium on VLSI circuits—VLSIC 2004* (pp. 176–179). New York: IEEE.
76. Lehmer, D. H. (1960). Teaching combinatorial tricks to a computer. In *Symposium on applied mathematics and combinatorial analysis* (pp. 179–193). Providence: AMS.
77. Lenstra, A. K., Hughes, J. P., Augier, M., Bos, J. W., Kleinjung, T., & Wachter, C. (2012). *Ron was wrong, Whit is right*. Cryptology ePrint Archive, Report 2012/064.
78. Lim, D. (2004). *Extracting secret keys from integrated circuits*. M.S. Thesis, Massachusetts Institute of Technology (MIT), MA, USA.
79. Lim, D., Lee, J. W., Gassend, B., Suh, G. E., van Dijk, M., & Devadas, S. (2005). Extracting secret keys from integrated circuits. *IEEE Transactions on Very Large Scale Integration (VLSI) Systems, 13*(10), 1200–1205.
80. Lin, L., Holcomb, D., Krishnappa, D. K., Shabadi, P., & Burleson, W. (2010). Low-power sub-threshold design of secure physical unclonable functions. In *ACM/IEEE international symposium on low power electronics and design—ISLPED 2010* (pp. 43–48). New York: ACM.
81. Linnartz, J.-P., & Tuyls, P. (2003). New shielding functions to enhance privacy and prevent misuse of biometric templates. In *Lecture notes in computer science (LNCS): Vol. 2688. International conference on audio- and video-based biometric person authentication—AVBPA 2003* (pp. 393–402). Berlin: Springer.
82. Lofstrom, K., Daasch, W. R., & Taylor, D. (2000). IC identification circuit using device mismatch. In *IEEE international solid-state circuits conference—ISSCC 2000* (pp. 372–373). New York: IEEE.
83. Maes, R., & Verbauwhede, I. (2010). Physically unclonable functions: a study on the state of the art and future research directions. In A.-R. Sadeghi & D. Naccache (Eds.), *Information security and cryptography. Towards hardware-intrinsic security* (pp. 3–37). Berlin: Springer.
84. Maes, R., Tuyls, P., & Verbauwhede, I. (2008). Intrinsic PUFs from flip-flops on reconfigurable devices. In *Benelux workshop on information and system security—WISSec 2008*. New York: IEEE.
85. Maes, R., Tuyls, P., & Verbauwhede, I. (2009). Low-overhead implementation of a soft decision helper data algorithm for SRAM PUFs. In *Lecture notes in computer science (LNCS): Vol. 5747. Workshop on cryptographic hardware and embedded systems—CHES 2009* (pp. 332–347). Berlin: Springer.
86. Maes, R., Tuyls, P., & Verbauwhede, I. (2009). Soft decision helper data algorithm for SRAM PUFs. In *IEEE international symposium on information theory—ISIT 2009* (pp. 2101–2105). New York: IEEE.
87. Maes, R., Van Herrewege, A., & Verbauwhede, I. (2012). PUFKY: a fully functional PUF-based cryptographic key generator. In *Lecture notes in computer science (LNCS): Vol. 7428. Workshop on cryptographic hardware and embedded systems—CHES 2012*. Berlin: Springer.
88. MagneTek(R). MagnePrint(R). http://www.magneprint.com/.
89. Maiti, A., & Schaumont, P. (2009). Improving the quality of a physical unclonable function using configurable ring oscillators. In *International conference on field programmable logic and applications—FPL 2009* (pp. 703–707). New York: IEEE.
90. Maiti, A., & Schaumont, P. (2011). Improved ring oscillator PUF: an FPGA-friendly secure primitive. *Journal of Cryptology, 24*, 375–397.
91. Maiti, A., Casarona, J., McHale, L., & Schaumont, P. (2010). A large scale characterization of RO-PUF. In *IEEE international symposium on hardware-oriented security and trust—HOST 2010* (pp. 94–99). New York: IEEE.

92. Maiti, A., Kim, I., & Schaumont, P. (2012). A robust physical unclonable function with enhanced challenge-response set. *IEEE Transactions on Information Forensics and Security*, 7(1), 333–345.

93. Majzoobi, M., Koushanfar, F., & Potkonjak, M. (2008). Testing techniques for hardware security. In *IEEE international test conference—ITC 2008* (pp. 1–10). New York: IEEE.

94. Majzoobi, M., Koushanfar, F., & Potkonjak, M. (2009). Techniques for design and implementation of secure reconfigurable PUFs. *ACM Transactions on Reconfigurable Technology and Systems*, 2(1), 1–33.

95. Massey, J. (1969). Shift-register synthesis and BCH decoding. *IEEE Transactions on Information Theory*, 15(1), 122–127.

96. Menezes, A. J., Vanstone, S. A., & Van Oorschot, P. C. (1996). *Handbook of applied cryptography*. Boca Raton: CRC Press.

97. Merli, D., Schuster, D., Stumpf, F., & Sigl, G. (2011). Side-channel analysis of PUFs and fuzzy extractors. In *Lecture notes in computer science (LNCS): Vol. 6740. International conference on trust and trustworthy computing—TRUST 2011* (pp. 33–47). Berlin: Springer.

98. Microchip Technology Inc. (2007). Serial Peripheral Interface (SPI). http://ww1.microchip.com/downloads/en/DeviceDoc/39699b.pdf.

99. Mitchell, T. M. (1997). *Machine learning*. New York: McGraw-Hill.

100. Morozov, S., Maiti, A., & Schaumont, P. (2010). An analysis of delay based PUF implementations on FPGA. In *Lecture notes in computer science (LNCS): Vol. 5992. International workshop on applied reconfigurable computing—ARC 2010* (pp. 382–387). Berlin: Springer.

101. Nisan, N., & Zuckerman, D. (1996). Randomness is linear in space. *Journal of Computer and System Sciences*, 52(1), 43–52.

102. Öztürk, E., Hammouri, G., & Sunar, B. (2008). Physical unclonable function with tristate buffers. In *IEEE international symposium on circuits and systems—ISCAS 2008* (pp. 3194–3197). New York: IEEE.

103. Öztürk, E., Hammouri, G., & Sunar, B. (2008). Towards robust low cost authentication for pervasive devices. In *IEEE international conference on pervasive computing and communications—PERCOM 2008* (pp. 170–178). New York: IEEE.

104. Pappu, R. S. (2001). *Physical one-way functions*. Ph.D. Thesis, Massachusetts Institute of Technology (MIT), MA, USA.

105. Pappu, R. S., Recht, B., Taylor, J., & Gershenfeld, N. (2002). Physical one-way functions. *Science*, 297, 2026–2030.

106. Puntin, D., Stanzione, S., & Iannaccone, G. (2008). CMOS unclonable system for secure authentication based on device variability. In *European solid-state circuits conference—ESSCIRC 2008* (pp. 130–133). New York: IEEE.

107. Quisquater, J.-J., & Samyde, D. (2001). ElectroMagnetic analysis (EMA): measures and counter-measures for smart cards. In *Lecture notes in computer science (LNCS): Vol. 2140. International conference on research in smart cards—E-SMART 2001* (pp. 200–210). Berlin: Springer.

108. Ranasinghe, D. C., Engels, D. W., & Cole, P. H. (2004). Security and privacy: modest proposals for low-cost RFID systems. In *Auto-ID labs research workshop 2004*. New York: IEEE.

109. Rényi, A. (1960). On measures of information and entropy. In *Berkeley symposium on mathematics, statistics and probability 1960* (pp. 547–561). New York: IEEE.

110. Research on Physical Unclonable Functions (PUFs) at SES Lab, Virginia Tech. (2012). http://rijndael.ece.vt.edu/puf.

111. Rivest, R. L., Shamir, A., & Adleman, L. (1978). A method for obtaining digital signatures and public-key cryptosystems. *Communications of the ACM*, 21(2), 120–126.

112. Rührmair, U. (2009). *SIMPL systems: on a public key variant of physical unclonable functions*. Cryptology ePrint Archive, Report 2009/255.

113. Rührmair, U. (2011). SIMPL systems, or: can we design cryptographic hardware without secret key information? In *Lecture notes in computer science (LNCS): Vol. 6543. Conference*

on current trends in theory and practice of computer science—SOFSEM 2011 (pp. 26–45). Berlin: Springer.

114. Rührmair, U., Chen, Q., Lugli, P., Schlichtmann, U., & Martin Stutzmann, G. C. (2009). *Towards electrical, integrated implementations of SIMPL systems*. Cryptology ePrint Archive, Report 2009/278.

115. Rührmair, U., Sölter, J., & Sehnke, F. (2009). *On the foundations of physical unclonable functions*. Cryptology ePrint Archive, Report 2009/277.

116. Rührmair, U., Busch, H., & Katzenbeisser, S. (2010). Strong PUFs: models, constructions, and security proofs. In A.-R. Sadeghi & D. Naccache (Eds.), *Towards hardware-intrinsic security* (pp. 79–96). Berlin: Springer.

117. Rührmair, U., Jaeger, C., Hilgers, C., Algasinger, M., Csaba, G., & Stutzmann, M. (2010). Security applications of diodes with unique current-voltage characteristics. In *Lecture notes in computer science (LNCS): Vol. 6052. International conference on financial cryptography and data security—FC 2010* (pp. 328–335). Berlin: Springer.

118. Rührmair, U., Sehnke, F., Sölter, J., Dror, G., Devadas, S., & Schmidhuber, J. (2010). Modeling attacks on physical unclonable functions. In *ACM conference on computer and communications security—CCS 2010* (pp. 237–249). New York: ACM.

119. Rührmair, U., Sehnke, F., Sölter, J., Dror, G., Devadas, S., & Schmidhuber, J. (2010). *Modeling attacks on physical unclonable functions*. Cryptology ePrint Archive, Report 2010/251.

120. Rührmair, U., Jaeger, C., & Algasinger, M. (2011). An attack on PUF-based session key exchange and a hardware-based countermeasure: erasable PUFs. In *Lecture notes in computer science (LNCS): Vol. 7035. International conference on financial cryptography and data security—FC 2012* (pp. 190–204). Berlin: Springer.

121. Rührmair, U., Jaeger, C., Bator, M., Stutzmann, M., Lugli, P., & Csaba, G. (2011). Applications of high-capacity crossbar memories in cryptography. *IEEE Transactions on Nanotechnology, 10*(3), 489–498.

122. Schindler, W., & Killmann, W. (2002). Evaluation criteria for true (physical) random number generators used in cryptographic applications. In *Lecture notes in computer science (LNCS): Vol. 2523. Workshop on cryptographic hardware and embedded systems—CHES 2002* (pp. 431–449). Berlin: Springer.

123. Schnabl, G., & Bossert, M. (1995). Soft-decision decoding of Reed-Muller codes as generalized multiple concatenated codes. *IEEE Transactions on Information Theory, 41*(1), 304–308.

124. Schrijen, G.-J., & van der Leest, V. (2012). Comparative analysis of SRAM memories used as PUF primitives. In *Design, automation and test in Europe—DATE 2012* (pp. 1319–1324). New York: IEEE.

125. Sedgewick, R. (1977). Permutation generation methods. *ACM Computing Surveys, 9*(2), 137–164.

126. Selimis, G. N., Konijnenburg, M., Ashouei, M., Huisken, J., de Groot, H., van der Leest, V., Schrijen, G. J., van Hulst, M., & Tuyls, P. (2011). Evaluation of 90 nm 6T-SRAM as physical unclonable function for secure key generation in wireless sensor nodes. In *IEEE international symposium on circuits and systems—ISCAS 2011* (pp. 567–570). New York: IEEE.

127. Shannon, C. E. (1948). A mathematical theory of communication. *Bell Systems Technical Journal, 27*, 623–656.

128. Sharma, A., Subramanian, L., & Brewer, E. A. (2011). PaperSpeckle: microscopic fingerprinting of paper. In *ACM conference on computer and communications security—CCS 2011* (pp. 99–110). New York: ACM.

129. Shimizu, K., Suzuki, D., & Kasuya, T. (2012). Glitch PUF: extracting information from usually unwanted glitches. *IEICE Transactions on Fundamentals of Electronics, Communications and Computer Sciences, E95.A*(1), 223–233.

130. Silverman, R., & Balser, M. (1954). Coding for constant-data-rate systems-part I. A new error-correcting code. *Proceedings of the IRE, 42*(9), 1428–1435.

131. Simmons, G. (1991). Identification of data, devices, documents and individuals. In *IEEE international Carnahan conference on security technology—ICCST 1991* (pp. 197–218). New York: IEEE.

132. Simons, P., van der Sluis, E., & van der Leest, V. (2012). Buskeeper PUFs, a promising alternative to D flip-flop PUFs. In *IEEE international symposium on hardware-oriented security and trust—HOST 2012* (pp. 7–12). New York: IEEE.

133. Škorić, B., Tuyls, P., & Ophey, W. (2005). Robust key extraction from physical uncloneable functions. In *Lecture notes in computer science (LNCS): Vol. 3531. International conference on applied cryptography and network security—ACNS 2005* (pp. 407–422). Berlin: Springer.

134. Škorić, B., Maubach, S., Kevenaar, T., & Tuyls, P. (2006). Information-theoretic analysis of capacitive physical unclonable functions. *Journal of Applied Physics, 100*, 2.

135. Su, Y., Holleman, J., & Otis, B. (2007). A 1.6 pJ/bit 96% stable chip-ID generating circuit using process variations. In *IEEE international solid-state circuits conference—ISSCC 2007* (pp. 406–611). New York: IEEE.

136. Suh, G. E., & Devadas, S. (2007). Physical unclonable functions for device authentication and secret key generation. In *Design automation conference—DAC 2007* (pp. 9–14). New York: ACM.

137. Suzuki, D., & Shimizu, K. (2010). The glitch PUF: a new delay-PUF architecture exploiting glitch shapes. In *Lecture notes in computer science (LNCS): Vol. 6225. Workshop on cryptographic hardware and embedded systems—CHES 2010* (pp. 366–382). Berlin: Springer.

138. Tarnovsky, C. (2010). Deconstructing a 'Secure' processor. Talk at Black Hat Federal 2010. http://www.blackhat.com/presentations/bh-dc-10/Tarnovsky_Chris/BlackHat-DC-2010-Tarnovsky-DASP-slides.pdf.

139. The Complexity Zoo (2012). http://qwiki.stanford.edu/index.php/Complexity_Zoo.

140. The SHA-3 Zoo (2012). http://ehash.iaik.tugraz.at/wiki/The_SHA-3_Zoo.

141. Tiri, K., Hwang, D., Hodjat, A., Lai, B., Yang, S., Schaumont, P., & Verbauwhede, I. (2005). Prototype IC with WDDL and differential routing—DPA resistance assessment. In *Lecture notes in computer science (LNCS): Vol. 3659. Workshop on cryptographic hardware and embedded systems—CHES 2005* (pp. 354–365). Berlin: Springer.

142. Tolk, K. (1992). *Reflective particle technology for identification of critical components* (Technical Report SAND-92-1676C). Albuquerque, NM, USA: Sandia National Labs.

143. Torrance, R., & James, D. (2009). The state-of-the-art in IC reverse engineering. In *Lecture notes in computer science (LNCS): Vol. 5747. Workshop on cryptographic hardware and embedded systems—CHES 2009* (pp. 363–381). Berlin: Springer.

144. Tuyls, P., & Batina, L. (2006). RFID-tags for anti-counterfeiting. In *Lecture notes in computer science (LNCS): Vol. 3860. Topics in cryptology: cryptographers' track of the RSA conference—CT-RSA 2006* (pp. 115–131). Berlin: Springer.

145. Tuyls, P., & Škorić, B. (2006). Physical unclonable functions for enhanced security of tokens and tags. In *Information security solutions Europe—ISSE 2006* (pp. 30–37). Wiesbaden: Vieweg.

146. Tuyls, P., Škorić, B., Stallinga, S., Akkermans, A. H. M., & Ophey, W. (2005). Information-theoretic security analysis of physical uncloneable functions. In *Lecture notes in computer science (LNCS): Vol. 3570. International conference on financial cryptography and data security—FC 2005* (pp. 141–155). Berlin: Springer.

147. Tuyls, P., Schrijen, G.-J., Škorić, B., van Geloven, J., Verhaegh, N., & Wolters, R. (2006). Read-proof hardware from protective coatings. In *Lecture notes in computer science (LNCS): Vol. 4249. Workshop on cryptographic hardware and embedded systems—CHES 2006* (pp. 369–383). Berlin: Springer.

148. Unicate, B. V. (1999). Guaranteeing identity and secure payments—the foundation for eCommerce. http://www.andreae.com/Unicate/emerge.pdf.

149. UNIQUE: Foundations for Forgery-Resistant Security Hardware (EC-FP7: 238811) (2012). http://unique-security.eu/.

150. van der Leest, V., Schrijen, G.-J., Handschuh, H., & Tuyls, P. (2010). Hardware intrinsic security from D flip-flops. In *ACM workshop on scalable trusted computing—STC 2010* (pp. 53–62). New York: ACM.
151. Van Herrewege, A., Katzenbeisser, S., Maes, R., Peeters, R., Sadeghi, A.-R., Verbauwhede, I., & Wachsmann, C. (2012). Reverse fuzzy extractors: enabling lightweight mutual authentication for PUF-enabled RFIDs. In *Lecture notes in computer science (LNCS): Vol. 7397. International conference on financial cryptography and data security—FC 2012.* Berlin: Springer.
152. Viterbi, A. (1967). Error bounds for convolutional codes and an asymptotically optimum decoding algorithm. *IEEE Transactions on Information Theory, 13*(2), 260–269.
153. von Neumann, J. (1951). Various techniques used in connection with random digits. *Journal of Research of the National Bureau of Standards, 12,* 36–38.
154. Vrijaldenhoven, S. (2005). *Acoustical physical uncloneable functions.* M.S. Thesis, Technische Universiteit Eindhoven, The Netherlands.
155. Yamakoshi, M., Tanaka, J., Furuie, M., Hirabayashi, M., & Matsumoto, T. (2008). Individuality evaluation for paper based artifact-metrics using transmitted light image. In *Conference series of the society of photo-optical instrumentation engineers (SPIE): Vol. 6819. Security, forensics, steganography, and watermarking of multimedia contents X.* New York: IEEE.
156. Yamamoto, D., Sakiyama, K., Iwamoto, M., Ohta, K., Ochiai, T., Takenaka, M., & Itoh, K. (2011). Uniqueness enhancement of PUF responses based on the locations of random outputting RS latches. In *Lecture notes in computer science (LNCS): Vol. 6917. Workshop on cryptographic hardware and embedded systems—CHES 2011* (pp. 390–406). Berlin: Springer.
157. Yin, C.-E. D., & Qu, G. (2010). LISA: maximizing RO PUF's secret extraction. In *IEEE international symposium on hardware-oriented security and trust—HOST 2010* (pp. 100–105). New York: IEEE.
158. Yu, M.-D. M., M'Raihi, D., Sowell, R., & Devadas, S. (2011). Lightweight and secure PUF key storage using limits of machine learning. In *Lecture notes in computer science (LNCS): Vol. 6917. Workshop on cryptographic hardware and embedded systems—CHES 2011* (pp. 358–373). Berlin: Springer.

Printed in the United States
By Bookmasters